ECO-URBANITY

The seemingly unchecked growth of cities in our increasingly globalized world can all too easily be seen as the cause of the ecological crisis that looms over us all. Urban areas, with their massive consumption of scarce resources and contribution to pollution and environmental damage, seem to be the problem, not the solution.

And yet this book argues that that is exactly what cities are – the victims of unsustainable growth and not the cause, the possible solution to global environmental pressures and not the problem. For it is cities that have always been the source of creative answers to the problems that humanity faces, and it is cities that can provide the synergy between creativity and ecology that the current situation demands.

Bringing together leading professionals and thinkers from across urban planning, design, landscape and architecture, *eco-urbanity* breaks down the barriers that so often stand between problems and their solutions. Encouraging innovative, creative thinking and the flow of ideas between theory and real-life experience, this book sets a new benchmark for how to make our future cities truly sustainable.

Darko Radović teaches architecture and urban design at the University of Melbourne. He has taught, researched and practised architecture and urbanism in Europe, Australia and Asia. *eco-urbanity* was conceived during his professorship at the University of Tokyo's Department of Engineering Centre for Sustainable Urban Regeneration.

ECO-URBANITY

Towards well-mannered built environments

EDITED BY DARKO RADOVIĆ

Routledge
Taylor & Francis Group

LONDON AND NEW YORK

First published 2009
by Routledge
2 Park Square, Milton Park, Abingdon, Oxon OX14 4RN

Simultaneously published in the USA and Canada
by Routledge
270 Madison Ave, New York, NY 10016, USA

Routledge is an imprint of the Taylor & Francis Group, an informa business

© 2009 selection and editorial matter, Darko Radović; individual chapters, the contributors

Typeset in Sabon by Wearset Ltd, Boldon, Tyne and Wear
Printed and bound in Great Britain by MPG Books Group, UK

All rights reserved. No part of this book may be reprinted or reproduced or utilized in any form or by any electronic, mechanical, or other means, now known or hereafter invented, including photocopying and recording, or in any information storage or retrieval system, without permission in writing from the publishers.

British Library Cataloguing in Publication Data
A catalogue record for this book is available from the British Library

Library of Congress Cataloging in Publication Data
Eco-urbanity: towards well-mannered built environments/edited by Darko Radović.
p. cm.
Includes bibliographical references and index.
1. Urban ecology. 2. Sustainable development. I. Radović, Darko.
HT241.E39 2009
307.76–dc22
2008040960

ISBN10: 0-415-47277-6 (hbk)
ISBN10: 0-415-47278-4 (pbk)

ISBN13: 978-0-415-47277-7 (hbk)
ISBN13: 978-0-415-47278-4 (pbk)

CONTENTS

Notes on contributors   vii
Acknowledgements   xiii

Introduction: towards a theory of eco-urbanity   1
DARKO RADOVIĆ

PART I
The compact city, strategies and success stories   7

1 Eco-urbanity: the framework of an idea   9
   DARKO RADOVIĆ

2 The Barcelona Agenda: reuse, compactness and green   19
   ORIOL CLOS

3 From industrial cities to eco-urbanity: the Melbourne case study   33
   ROB ADAMS

4 The sustainable city as a fine-grained city   47
   DAVID SIM

5 From the compact city to the defragmented city: another route towards a sustainable urban form?   63
   MIKE JENKS

PART II
Other cultures, approaches and strategies   75

6 Designing for shrinkage: Fibercity 2050, Tokyo   79
   OHNO HIDETOSHI

7 Excavating the lost commons: creating green spaces and water corridors for eco-urban infrastructure   92
   ISHIKAWA MIKIKO

8 Continuity and departure: a case study of Singapore's Nankin Street 103
HENG CHYE KIANG

9 The cultural challenge for sustainable cities: coping with sprawl in Bangkok and Melbourne 112
SIDH SINTUSINGHA

10 Geometries of life and formlessness: the theoretical legacies of historical Beijing 125
JIANFEI ZHU

11 Eco-city? eco-urbanity? 141
ARVIND KRISHAN

PART III
Other scales and sensibilities 161

12 Eco-urbanism: an Israeli perspective 163
ARIE RAHAMIMOFF

13 Bringing back nature and re-invigorating the city centre 176
KENGO KUMA

14 Sustainable design towards a positive spiral 178
KODAMA YUICHIRO

15 Creating a cemetery: architecture that sustains cultural forms 182
NAITO HIROSHI

16 *eco-urbanity* hypothesis: towards well-mannered built environments 187
DARKO RADOVIĆ

Index 192

CONTRIBUTORS

Rob Adams is one of the champions of urban design in Australia, with over 30 years of experience as a practising designer in both private enterprise and the public service. As a Director of City Design, he is one of six directors at the City of Melbourne. In 1985 Rob put in place the first comprehensive urban design strategy for the City of Melbourne. This programme has been successfully implemented over the last two decades. The amazing evolution of central Melbourne has been well-documented and publicized (e.g. in *Places for People*, a book produced by the City of Melbourne in association with Jan Gehl). More recently Rob and his office have made significant contributions to the City's 'Zero Emissions 2020' project, through such buildings as CH_2 (Australia's first six-star commercial building), the East Melbourne Library and the solar installations at the Queen Victoria Market. For those and other projects his division has received over 70 state and national awards for design excellence. Rob has been a visiting lecturer at RMIT and Melbourne University. In June 2004 the University of Melbourne recognized his personal contribution to the University and the city by giving him the title of Professorial Fellow. He is a regular keynote speaker at national and international conferences and has written numerous papers and articles.

Oriol Clos is the Chief Architect of the City Council of Barcelona. He has also served as Director of Urbanism (Project 22@bcn) and as Director of Plans and Urban Projects for the City Council of Barcelona. In the 1990s and early 2000s Oriol was Associate Professor at the Department of Urbanism, School of Architecture of Barcelona. He also led his private practice, working on a wide range of building and planning projects that include the enlargement of Parliament in Madrid, facilities and public spaces for the Olympic Games in the diagonal area of Barcelona and in the Olympic Village of Banyoles, the Salvador Dalí Museum in Púbol i Cadaqués, the Ville-Port sector in Saint-Nazaire, France (with Manuel de Solà-Morales) and urban projects in the new diagonal sector and Poblenou, Barcelona. Oriol's research focuses on the strategic recycling of green spaces and urban compactness of central urban industrial areas.

Heng Chye Kiang is a professor and Dean of the School of Design and Environment at the National University of Singapore and a co-leader of the Asian cities cluster at the Asian Research Institute. He studied architecture at the

École spéciale d'architecture, and urban design at the École nationale des travaux publics de l'état. He did his PhD at the University of California. Heng Chye Kiang has been a visiting scholar at Tsinghua University and Kyoto University, and Huaying Visiting Professor at the Southeast University in Nanjing. He has served as a jury member in a number of international design competitions as well as on several editorial boards of international journals. He is a board member of Singapore's Urban Redevelopment Authority. Heng Chye Kiang's research covers areas of urban design, and the heritage and history of Chinese cities. He has published widely in these areas and was awarded the NUS Outstanding Researcher Award in 1997. His book, *Cities of Aristocrats and Bureaucrats* (University of Hawaii, 1999) is used as a textbook in leading universities in the United States and the United Kingdom. China's Architecture and Building Press has recently published his digital reconstruction of the Chang'an Tang period as a book and an interactive software package.

Ishikawa Mikiko is a landscape architect and city planner. She is Professor of Environmental Design at the Faculty of Environmental Information, Keio University and was a Visiting Scholar at the Graduate School of Design, Harvard University from 2004–5. She is the author of a number of articles and books, including *Theory and Practice of Water Basin Environmental Management and Design* (2005), *Perspectives of Public Spaces in Urban Planning* (2005), *Revitalization Cities based on the Reformation of Natural Oriented Watershed Basin* (2005), *Design for Sustainable Community* (2004) and *City and Green Space* (2001). Her recent major landscape design and planning works include the historical park of Shari Tower in Shenyang (2007), the regeneration of the Hun River in Shenyang (2007), the Citizen Forest in Kakamigahra (2006), the Meditation Forest (2005), a crematory (in collaboration with Toyo Ito), the Ecological Park in Kawashima (2005), and the Miyama Wetland Park (2005).

Mike Jenks is Professor Emeritus in the Department of Architecture, Oxford Brookes University. He is the founding Director of the Oxford Institute for Sustainable Development. Currently, as principal investigator, he leads a large national research group called City Form: the Sustainable Urban Form Consortium, involving five universities and 14 non-academic organizations. City Form is funded by the Engineering and Physical Sciences Research Council. Recently he won a grant from the Higher Education Funding Council to act as an expert independent evaluator to assess a large project building sustainable communities. Mike's research expertise relates to the compact city and sustainable urban form, the intensification of development in urban areas, design guidance, housing and housing layout, and the influence of layout design on safety and security. This expertise has been in considerable demand worldwide, and most recently he has found himself advising metropolitan governments in Japan, Taiwan and Thailand. Mike's co-edited publications include an influential trilogy of books, *The Compact City: A Sustainable Urban Form?* (1996), *Achieving Sustainable Urban Form* (2000) and *Compact Cities: Sustainable Urban Forms for Developing Coun-*

tries (2000), all published by Spon Press. He co-wrote with Nicola Dempsey, *Future Forms and Design for Sustainable Cities* (2005), published by the Architectural Press. His latest book, *World Cities and Urban Form: Fragmented, Polycentric, Sustainable?* written with Daniel Kozak and Pattaranan Takkanon, was published by Routledge in 2008.

Kodama Yuichiro has long been engaged in research and design practices on passive heating and cooling systems at the Building Research Institute of the Ministry of Construction, Government of Japan. He is a well-known developer of interactive design tools for passive design and for the life-cycle analysis of buildings. He is an author of prominent books on bioclimatic design, passive design and sustainable design. He is also well-known as a pioneer architect in this field and has been awarded the Architectural Institute of Japan Architectural Design Commendation, and a Japanese Good Design Award. His international activities include the Passive and Low Energy Architecture Network, the International Tropical Architecture Network and IEA energy conservation in building and community systems programme. Since 1998 he has been a professor at Kobe Design University.

Arvind Krishan is an architect, planner and structural engineer, with a Masters in architecture from the University of Washington, a Masters in structural engineering and a PhD in energy-efficient architecture from the Indian Institute of Technology, Delhi. He is a senior professor at the Department of Architecture, School of Planning and Architecture, New Delhi and has been the Dean of Studies and the Head of the Department of Architecture at this school. Arvind is a practising architect in the fields of climate-responsive architecture. He has designed buildings of various sizes and programmes in India and abroad. His design is supported by research and funded through various national and international agencies such as the British Council, the European Union and the Government of India. He is an international consultant to the World Bank for their projects in China and a member of UNEP expert group on eco-cities. He has won design awards in competitions and publication awards for his research work, as well as lecturing extensively in India and abroad.

Kengo Kuma is an internationally known Japanese architect and a professor at the Keio University in Tokyo. He is the founder of Kengo Kuma & Associates, established in 1990. His approach to architecture has brought him numerous international awards, including the American Institute of Architects' Dupont Benedictus Award for Water/Glass in 1997, the International Stone Architecture Award in Italy in 2001 for his Stone Museum, and first prize at the Spirit of Nature Wood Architecture Award, in Finland in 2002. Kuma's portfolio includes an abundance of works from bathhouses to museums, commercial, residential and university complexes, retail, mixed-use spaces and hotels. He is also the author of a number of books and articles. Kuma's design philosophy focuses on the relationship between architecture and nature, and the relationship between materials. At the core of his approach is a desire to blend architecture with nature, the use of natural materials and ceaseless efforts to create well-ventilated spaces filled with sunlight.

Naito Hiroshi is an architect, a founder of Naito Architects and Associates and a professor at the Department of Civil Engineering, University of Tokyo. He is the recipient of many design awards and his work has been exhibited in many countries around the world. Recent projects include Garaku Annex, Toyama Prefecture and the Shimane Arts Centre, Shimane Prefecture in 2005, the Rinri Institute of Ethics, Funabashi Housing, Chiba Prefecture, Tomata Dam Control Centre, Okayama Prefecture and the Minato-mirai Line, Yokohama, Kanagawa Prefecture. Naito Hiroshi's current research focuses on cemeteries. In parallel to this, his office is involved in several designs of cemeteries.

Ohno Hidetoshi is a professor at the Department of Socio-Cultural Studies, Graduate School of Frontier Science, University of Tokyo. He has been a Visiting Professor at the Department of Architecture, St Lucas Higher Institute, Brussels. He has lectured extensively in Europe and Asia. His current research focuses on use of urban fibres to develop an alternative model of the metropolis in an era of shrinking cities. His design projects are done in partnership with the Architect and Planners League, Tokyo. Ohno's most recent projects includes Nakagawa Primary School, Fukui Prefecture (2007), Freude Hikoshima, home for the elderly, Yamaguchi Prefecture (2005), Ukai Ohashi Bridge, Gifu Prefecture (2004) and the Heart Home, Hirakawa, a home for the elderly in Yamaguchi Prefecture (2003).

Darko Radović is a professor at the Centre for Sustainable Urban Regeneration of the University of Tokyo. He received his doctorate in architecture and urbanism from the University of Belgrade, Yugoslavia. He has taught, researched and practised architecture and urbanism in Europe, Australia and Asia. His recent research focuses on situations where architecture and urban design overlap, where traditional architectural and urban scales blur, where the social starts to acquire physical form. Darko's investigation of the concepts of urbanity and sustainable development focuses on culturally and environmentally diverse contexts, contexts that exemplify and expose difference and offer encounters with the other. Over the last three years Darko guest-lectured in France, Japan, Serbia and Montenegro, Singapore, Spain, Thailand and Vietnam. He has published in Europe, Australia and Asia. In 2005, his book *Green City* (co-written with Low, Gleeson and Green) was published by Routledge and UNSW Press. In 2007 the University of Belgrade published *Urbophilia* and Routledge published *Cross-cultural Urban Design* (co-written with Bull, Boontharm and Parin). While in 2008 the University of Tokyo published *Another Tokyo*.

Arie Rahamimoff is an architect and urban planner. He runs a private practice in Jerusalem, Israel, where he has worked since 1970. His most recent projects include the Ming Tombs, Beijing (a conservation master plan with Li Dezhong and Giora Solar, 2006–7), copper mines in southern Israel, Timna (a detailed master plan of 2,000 ha in one of the copper mines from

4000 BC for AHMSA), the Diaolou Towers and the villages of Kaiping, south China (conservation, economic, development and tourism, with Tan Weiqiang and Giora Solar for the State Administration of Cultural Heritage of the People's Republic of China), Birgu, Malta, master plan for the ancient capital of Malta (preservation, tourism, economic development for the Cottonera rehabilitation project, with Dr R. Bondin and Giora Solar) and with Coexistence, Cooperation, Partnership, a project for alternative futures for the region of Beit She'an (Israel), Jenin (Palestinian Authority), Northern Jordan, Harvard Graduate School of Design with Professor Carl Steinitz. Arie has published over 30 articles in international and Israeli journals and books. He has also been a guest lecturer at UNESCO–International Convention on Monuments and Sites international congress, Visiting Professor at the School of Architecture, University of New Mexico and for the University of Stuttgart, a visiting critic in urban design for the Graduate School of Design, Harvard University, a member of the District Committee of Jerusalem, a jury member in design competitions in Israel and abroad and recipient and winner of eight first prizes in architecture and urban design competitions.

David Sim received his Masters in architecture from the Edinburgh College of Art, Heriot-Watt University, Scotland. He is a senior lecturer in architecture and urban design at the Lund University, Sweden and an Associate at Gehl Architects Urban Quality Consultants, Copenhagen. David is well-known in Swedish professional circles as an educator. He has taught at architecture and design schools worldwide, while his work at Lund University has led to a number of pedagogical awards and distinctions. While in Lund David also researched and developed patterns for greater density and diversity in urban areas. This research has been successfully applied in urban design projects and competitions. David continues his work as an educator at Gehl Architects both with professionals and the public, giving lectures and guest critiques, as well as facilitating workshops and other planning events. His main area of work at Gehl Architects is in urban design, collaborating with other professionals in the planning and building process and applying Jan Gehl's theories to large-scale projects. These include master-planning new settlements in Ireland at Cherrywood, Kishoge and Clonburris and in Scotland at Camus Mor and Calderwood.

Sidh Sintusingha is a lecturer and coordinator for the bachelor of landscape architecture programme at the University of Melbourne. His research interests include cross-disciplinary approaches to urban sustainability, suburban sprawl in Asian cities, and the local negotiations with global expressions of both designers and laypeople, as manifested in Thai cultural landscapes. Sidh has lectured and taken part in conferences in Thailand, Australia, Vietnam and France. His most recent publications include: 'Sustainability and urban sprawl: alternative scenarios for a Bangkok superblock' and 'Sustaining sustainability and its inconsistencies'. He received honourable mention in the Ellis Stones Memorial Award for his article, 'Regionalism and the global suburb in Southeast Asia', co-written with David O'Brien and Phuong Dinh in 2005 and 2nd Dean's Prize for

Published Postgraduate Research in 2004 for his article, 'Bangkok: sustainable sprawl?' in *NaJua*, the journal of the Faculty of Architecture of Silpakom University.

Jianfei Zhu studied architecture at Tianjin University in China and obtained his PhD at the Bartlett School, University College, London. He is a senior lecturer at the University of Melbourne. His research explores modern architecture, issues of urbanity in China and East Asia, and east–west interactions in thinking and practice. He is the author of *Architecture of Modern China: A Historical Critique* (Routledge, 2008) and *Xingshi yu Zhengzhi* ('Form and politics: studies in Chinese architecture', manuscript in progress), *Didu Beijing de Kongjian Celue 1420–1911* ('Spatial strategies of imperial Beijing 1420–1911'). He has also written a number of articles for various journals. Over the last few years Jianfei has been guest lecturer in Italy, Canada, the United Kingdom, Australia and China and has made media appearances about his work on contemporary Chinese architecture.

ACKNOWLEDGEMENTS

From the very first moment, my *eco-urbanity* initiative received active support from a number of institutions and individuals, for which I owe my deepest gratitude. The project consisted of three phases: the preparations for the *eco-urbanity* Symposium, the actual Symposium, which was held at the University of Tokyo in September 2007, and the post-symposium collaboration on the production of this book.

First, I would like to thank the School of Engineering, Department of Urban Engineering and the Centre for Sustainable Urban Regeneration (cSUR) of the University of Tokyo for their generous support in organizing and hosting the *eco-urbanity* Symposium. My special thanks go to Professors Ohgaki Shinichiro and Okata Junichiro, who led the cSUR, the numerous cSUR researchers who, at various stages, assisted in the organization of the event and, most importantly, to the selfless Dr Nakamura Hitoshi, who was a dynamo behind the organization and without whose personal effort we would never have managed to stage the event. I also wish to thank a number of postgraduate students at the University of Tokyo, and especially those who became the creative force behind the production of all the Symposium material – Rosalinda Baez, Akiyoshi Inasaka, Wimonrart Issarathumnoon, Heide Jäger, Marieluise Jonas, Heike Rahman and Zhang Guangwei.

Then, my deepest gratitude goes to the participants of the Symposium and the authors of this volume: Rob Adams, Laretna Adishakti, Oriol Clos, Heng Chye Kiang, Ishikawa Mikiko, Mike Jenks, Jinnai Hidenobu, Kodama Yuichiro, Kengo Kuma, Naito Hiroshi, Ohno Hidetoshi, Arie Rahamimoff, David Sim, Sidh Sintusingha and Jianfei Zhu. Laretna Adishakti took part in the Symposium, but other commitments prevented her from contributing to this volume. Jinnai Hidenobu joined the Symposium to co-chair the event with me. His strong intellectual contribution reaches far beyond that event and has made a significant contribution to the ideas presented in the book.

In final phase, the production of the manuscript for this book, I divided my time between my work in Tokyo and Melbourne. The University of Tokyo supported the project to the very end, while the University of Melbourne came in with much-needed support during the production of the manuscript. In that phase, my special thanks go to the copy-editor, Carole Pearce, who made a huge effort to preserve the sense of cultural difference in the chapters that were written by authors of such different cultural backgrounds,

ACKNOWLEDGEMENTS

and to Mirjana Ristic, who assisted with the graphics and made sure that our material was up to high Routledge standards.

And, lastly, my thanks go to Routledge: Caroline Mallinder, for her encouragement and endorsement of a book that recognizes the need to visually present urban and architectural space, and to Georgina Johnson, Alex Hollingsworth and Allie Waite, who confidently guided our project to its conclusion.

D.R.

INTRODUCTION
Towards a theory of eco-urbanity

DARKO RADOVIĆ

I agree with Michael Sorkin that 'Cities are units of human accountability to the planet' (Sorkin 1993: 11). As sensitive reflections of broad societal values, trends and trajectories, cities reflect the highs and the lows of human civilization. Each city is a 'projection of society on the ground', and 'that is, not only on the actual site, but at a specific level, perceived and conceived by thought, which determines the city and the urban' (Lefebvre 1996: 103). The urban is where 'social relations ... project themselves into a space, becoming inscribed there, and in the process producing the space itself' (Lefebvre 1991 in Soja 1996: 46). Those inscriptions help us read the spatial textures of the city, its agoras, streets, squares, fortresses, markets, palaces and prisons, as texts with which we can decipher the social origins and ideological charge of certain historical epochs.

Built environments, thus, accurately record human achievements. Those records point at broader social issues and problems. In the context of the topic of this book, which is urban sustainability, we can simply say that, if our actions are based on values that respect the environment, then our cities will be sustainable: if our guiding social values are not sustainable, then our cities become unsustainable, too.

We live in an increasingly globalized world in which the continuation of dominant power structures relies on practices that are extremely damaging to both natural and social aspects of the environment. As the consequence, the quality of life in cities is deteriorating. Cities are the nodes of densification and intensity of human activities and their impact reaches far beyond their physical limits. A radically unsustainable society inscribes deep wounds into the urban fabric: wounds that are difficult to heal and that create scars that are impossible to hide.

The cities are not the villains of the piece. They are the victims of the profoundly unsustainable growth patterns that are dictated by global markets, the 'false ideological universality' of which perpetuate established power relations. These relations 'mask and legitimise a concrete politics of Western imperialism, military interventions and neo-colonialism' (Žižek 2005: 128).

This value system is not new. What is new is its dramatic acceleration and strength. It embodies a well-established, colonial perspective and is characterized by arrogance and a tendency towards the total domination of the other. It expresses global(ized) capital and promotes itself as an ideology to which there is no alternative. The founding myths of that ideology include the free market, a belief in non-differentiated progress, universal democracy, universal human rights and the unavoidability of dependence. The spatial projections of those myths and the associated growth patterns are cities and architecture that respect neither their ecological nor their cultural contexts and all look alike – the American CBDs of the rich, the *favelas* and shanty towns of the poor, and the huge, nondescript settlements of the middle class in between. Such cities are in a dire crisis, as much as our societies are. But cities remain the nodes of human creativity. It is within their creative potential that we need to search for paths towards better, sustainable ways of life.

That is the broadest credo lying behind this book.

The eco-urbanity initiative merges the two statements offered above. The first is that if we want sustainable cities, then we have to develop a profoundly different sociocultural framework in which such settlements can happen. The second is that only cities possess the creative power that can generate radically different ways of thinking, a much-needed paradigm shift. That shift calls for action against the powerful value system ruling us.

This book positions that imperative into the field of the production of space. The overarching questions addressed by the authors of *eco-urbanity* include: what current ideas and practices in the production of space have the potential to contribute to the much-needed change? Which of those ideas and practices may be recognized as steps in the right direction? Which of them constitutes an active contribution to the necessary paradigm shift?

In September 2007, during my professorship at the University of Tokyo, I invited a group of leading academics and practitioners of architecture, landscape architecture, urban design and urban planning from Australia, China, Denmark, India, Indonesia, Israel, Japan, Singapore, Spain, Thailand and the UK to Todai's Centre for Sustainable Urban Regeneration to participate in a symposium to discuss those questions. Our task was to cross common disciplinary and cultural boundaries, to think together and seek a better understanding of what constitutes sustainable practice today. We believed that such a practice should be capable of embracing both environmental responsibility and cultural responsiveness. It should be locally accountable and thus globally relevant. It should facilitate both thinking and acting in the direction of what we tentatively call eco-urbanity.

Focused as it is on urban and architectural projects and practices that successfully combine complex environmental and cultural responsiveness and responsibility, this book celebrates cultural difference. Cultural otherness and mutual respect were the key ingredients of the *eco-urbanity* Symposium that, I believe, is reflected in this volume. That includes the co-existence of a variety of ideas, locations, ways of thinking, approaches and even writing and presentation styles. While the language of this book is English, most of the ideas presented here come from other cultural frameworks. Assembled in this book, their combined message reaches far beyond the limits of any single cultural framework. The languages in which these ideas were conceived – Catalan, English, Spanish, Thai, Mandarin Chinese, Japanese and others – remained alive throughout the Symposium, and I hope that they will linger on between the lines of this book, to the extent of causing a certain friction and even discomfort. That would be the result of a conscious effort to hint at the complexity of our topic, the set of themes that reach far beyond our current knowledge, and stress the need for continuing communication and exchange of ideas among equals.

Languages express ideas. Ideas belong to places, times and cultures of their origin. They possess distinct idiolects and sociolects. We live in a world that increasingly tames difference. Built environments produced by global capital are the projection of global capitalism, expressing value systems that are often narrowly described as 'Western'. In the process, the local and the particular are lost, sacrificed to the global and general. As dominant power disseminates its preferred ways of thinking, such homogenization happens at all levels of our existence. Thus, 'non-Western cultural theory remains invisible' (Highmore 2002: 177). The same applies to urban planning, urban and landscape design and architectural theory, and this is happening at a time when civilization demands exactly the opposite. Donna Haraway (1991: 183) eloquently demands *situated knowledges*, thus celebrating both the place and the plurality of ways of knowing. ('Knowledges' is a subversive plural which, we are told, as we are told about many other alternatives, is impossible.) This book ambitiously aims to make a very small step on the path towards making a radical qualitative change in the ways we conceive, perceive and live our urban environments.

This introduction is followed by Chapter 1, which summarizes the seeds of the idea of eco-urbanity. The 14 chapters that follow are structured in three parts. They loosely follow the physical scales of the projects discussed by the contributors to the book. Those scales range from broad urban frames down to the finest of architectural details, and they open out a variety of themes and issues for discussion and further thinking. *eco-urbanity* opens up as many questions as it answers. In that sense, this work aims to be an *opera aperta*. It wants to provoke and engage the reader.

Part I, 'The compact city, strategies and success stories', offers examples of two cities that are often quoted as best practice in recent urban development: Barcelona and its ability to reinvent its urban self (Clos), and the exemplary transformation of inner Melbourne (Adams). Development in both of those cities confirms an established body of theory which identifies urban compactness with better future. Sim presents current thinking about the liveability and quality of urban space by stressing the importance of unmeasurable qualities of the urban. Mike Jenks, one of the key proponents of the idea of compact city follows with a much-needed word of caution. He reminds us that, although the idea of compact city may promote good development in certain parts of the world, it is not and cannot be a panacea. His argument leads to explorations of cultures and spaces profoundly different from those in the West.

The second part, 'Other cultures, approaches and strategies', offers examples of these cultures and spaces. It presents urban studies, projects, initiatives and spatial thinking from Asian cultures as diverse as Japan, Thailand, Singapore, China and India. Here cities shrink or uncontrollably expand. Some topics in this section are relevant globally, while others introduce values that are profoundly different from those glorified by the (stereo)typical West. Ohno speaks about the future of shrinking Tokyo. Ishikawa reports on ways of empowering local communities, while Heng presents another kind of urban regeneration in post-colonial Singapore. Sintusingha's chapter addresses the globally relevant problem of urban sprawl, using examples of two culturally distant cities – Bangkok and Melbourne. Jianfei Zhu's work explores the ways in which traditional spatial forms and practices can be re-energized and contribute to the efforts to localize production of space. Arvind Krishan presents a generous selection of urban planning and architecture to address different levels of spatial form in the variety of scales.

The third part deals with 'Other scales and sensibilities', with an emphasis on environmentally and culturally sustainable architecture. Rahamimoff presents the continuity of principles of environmentally responsive architecture and subtle cultural responsiveness in a number of examples from Israel. Kuma's largely visual chapter continues the theme of learning from the past, richly illustrated by his highly effective, decidedly contemporary projects where the overall schemes and the finest of detailing seek both continuity and cultural relevance. Kodama adds to Kuma's emphasis on the importance of multiscale thinking, and brings in the issue of interdisciplinarity and interconnectedness. Naito ends the third part with his strongly visual exploration of death, cemeteries and graves as the ultimate inscription of our existence on the ground.

Each of these three parts is followed by the colour insert. The inserts not only illustrate the texts presented in the preceding pages, but provide additional information and a visual continuation of those chapters, thus making another kind of text which invites the open, critical and creative involvement of the reader. The book concludes with a definition of the elements of an eco-urbanity hypothesis, the scaffold for a theory of environmentally and culturally appropriate urban space.

This book is decidedly multivocal. The places, the times, the names of the actors that constitute this project best explain the emphasis on diversity in this proudly heterogeneous product. Adams, Clos, Heng, Ishikawa, Jenks, Kodama, Krishnan, Kuma, Naito, Ohno, Radović, Rahamimoff, Sim, Sintusingha, Zhu. Australia (via Zimbabwe), Spain (Catalonia), Japan, UK, Australia (via Yugoslavia, a country that no longer exists), Israel, Denmark (via Sweden and Scotland), Singapore, China (Australia), India and Thailand.

The names of Japanese and Chinese contributors to this volume are written in the Japanese and Chinese way, with their family names first. The names of other authors, again, are written in the way that is common in their cultures – with personal names first. The editor's own culture allows for the name to be written in both ways. That, it is to be hoped, makes Darko Radović/Radović Darko well-suited for his role of mediator.

The terminology of the eco-urbanity dialogue is rich but it depends on some common keywords: ecology, environment, urbanity, other, difference, ethics, continuity, change, many, respect, responsibility, responsiveness, sustainability. The vagueness and multiple meanings of many of those terms are an important quality of a topic that remains open for exploration and further definition. From this common basis the terminology expands, exploding into a variety of idioms, jargons, languages and accents.

Each of the projects in this volume departs from a common basis, which is loosely framed by those keywords, and transformed into unique concepts that are then translated into concrete spaces. Those concepts and spaces resonate with a significant level of difference from the business-as-usual in thinking and production of space. It is to this difference that we want to point.

BIBLIOGRAPHY

Haraway, D. (1991) *Simians, Cyborgs and Women: The Reinvention of Nature*, London: Free Association.

Highmore, B. (2002) *Everyday Life and Cultural Theory*, London and New York: Routledge.

Lefebvre, H. (1996) *Writings on Cities*, Oxford and Cambridge, MA: Blackwell.

Soja, E.W. (1996) *Thirdspace, Journeys to Los Angeles and Other Real-and-Imagined Places*, Cambridge, MA and Oxford: Blackwell.

Sorkin, M. (1993) *Local Code; The Constitution of a City at 42 N Latitude*, New York: Princeton Architectural Press.

Žižek, S. (2005) 'Against human rights', *New Left Review* 34, July/August: 115–31.

Part I

THE COMPACT CITY, STRATEGIES AND SUCCESS STORIES

INTRODUCTION

The body of the book contains three clusters of chapters. The first cluster of four chapters deals with the idea of the compact city, the theories and strategies behind its conception and its implementation, and examples of best practice.

Barcelona is widely recognized as a powerful example of an urban transformation that has largely been achieved by utilizing its own social energies and cultural resources. Oriol Clos reminds us how, over the last 25 years, as land available for urban growth became increasingly scarce, the city has been working on re-energizing precincts that have lost their original role. Following its own development concept of urban implosion, Barcelona is recycling the central parts of its urban fabric that were initially conceived for highly specific, mainly industrial use. As a hands-on urban practitioner and the urban manager of Barcelona, Oriol Clos focuses on a number of concrete issues. These include land use, intensity, new forms of urban fabric, an understanding of public space as a type of infrastructure, the celebration of open space, public housing and the value of productive and tertiary activities, together with their capacity to replace the traditional industrial activities that are now no longer found in the city centres. Those measures summarize the efforts of the city of Barcelona to weave a strong environmental concern into the fabric of a historically established urbanity.

Rob Adams argues that the twentieth century will go down in history as the century that lost the art of city-making. He believes that the twenty-first century is already seeing the convergence of new urban design approach, with special considerations of future urban sustainability. For Adams, factors such as density, mixed use, connectivity, adaptability, local character and a high quality of public realm, which are recognized as driving urban liveability, are identical to those that are responsible for environmental sustainability. The road to eco-urbanity, Adams concludes, is likely to parallel the roads to liveability and economic vitality. Working along those lines, Melbourne has transformed its city centre district from a mono-functional central business district to a multifunctional activity centre that has, on numerous occasions, won the world's most liveable city rating. Adams' chapter presents the strategies implemented in Melbourne over the last 20 years to create a more liveable and now a more sustainable city. The chapter outlines future projects that could propel Melbourne to becoming exemplary for its eco-urbanity.

The theme of liveability is further elaborated by David Sim. He draws our attention towards the importance of thick thresholds between public and private realms. He refers to damaging aspects of globalization, the ever-increasing competition between cities and the resulting developments that often share planning principles and a sterile corporate look. This trend towards uniform and naked urban environment leads to cities that are characterized by sharp distinctions between the public and the private. The opposite is true of the traditional city, whether in Japan or in Denmark, which grew up on the basis of a variety of principles that shape the spectrum from wholly public to wholly private space, thus permitting a range of a lively and locally rooted urban experiences to its citizens. David Sim suggests that a sustainable city has to be a liveable city where people feel comfortable. That is why planning must take into account a number of human needs, including the need for different degrees of intimacy. The argument is that we need to develop methods that will reintroduce the varying grades of intimacy and access between the public and the private. Qualitative, experiential, tactile and sensual aspects of urbanity may be hard to measure, but that does not make them less important than quantitative aspects.

Mike Jenks, who has, over the years, made an important contribution to the idea of compact city, offers in his chapter a succinct summary of the idea of urban compactness in order to ask whether the notion that that approach offers sustainable solutions for all parts of the world is justified. He argues that the practice has, to an extent, overtaken the knowledge and evidence needed to assure its success, and suggests that the largely Eurocentric theory may have exceeded its capacity for replication. It cannot simply be scaled up to fit a context of fast-growing megacities and megacity regions. At these scales cities tend to move from traditional monocentric forms towards polycentric structures, which often lead to undesirable urban fragmentation. A new paradigm is needed that is both capable of incorporating relevant aspects of compact-city theory and is physically and culturally meaningful in a non-Western context. This may require a process of integration, of bringing together urban, social and economic fragments – of defragmenting the city. This chapter provides a reminder that ideas, including those about sustainable development, belong to particular cultures, and shows how their uncritical global implementation and reduction to simplistic rules can be counter-productive and damaging.

1 ECO-URBANITY
The framework of an idea

DARKO RADOVIĆ

THE CRISIS

When we discuss sustainable urbanism and architecture, we usually speak about the principles of *ecological* sustainability. A rigorous translation of those principles into action is of critical importance, not only for the production of better built environments, but also for the planet at large.

For decades we have been aware of the galloping environmental crisis. A number of books produced in response to the oil crisis documented the depth of the problem in the 1970s. The origins and the nature of crisis were well-argued almost four decades ago. The seminal book, *Limits to Growth* (Meadows *et al.* 1972) was followed by Erich Fromm's convincing arguments in *To Have or To Be* (1976), Schumacher's *Small Is Beautiful* (1993), Lovelock's *Gaia Hypothesis* (1979), Bookchin's ideas in *Towards an Ecological Society* (1980), to name only the best-known titles, all of which confirmed the state of crisis and demanded urgent action. However, for all sorts of political and ideological reasons, those calls remained unanswered. Action was delayed and the very possibility of change severely compromised. Today, while we can lament how relevant those old books remain, we are sorely aware that we have to deal with a situation that is much worse than it was in the 1970s.

Even then, the state of the environment was alarming, but the ruling form of capitalism was uncomfortable with the very notion of crisis. After the debacles of Vietnam War and the 1968 unrests, the perception of stability, rather than further disturbances, was an imperative for those in power. Environmentalist agendas included strong questioning of emergent neo-liberal thinking. Their values directly confronted the world order and its well-recognized greed, individualism and materialism. That rendered them subversive and thus 'unrealistic'. The reaction was swift and to a large degree it targeted the core issue of the dissent – the philosophical values on which it was based. The post-1968 *ethical turn*, which grew from the ashes of the 1968 movements, promoted the fragmentation of dissent and the individualization of ethics, to make them compatible with and acceptable to the dominant conservative world-views (Bourg 2007: 32). As Kristin Ross noticed, that was 'a flight from genuine politics to cultural questions and personal meaning' (Bourg 2007: 12). It was that turn, rather than the much-needed turn towards a more just and sustainable world, that defined the fate of environmentalist and other broad social initiatives over the decades that followed. The movements with any capacity to introduce true, structural societal change were gradually dismantled, fragmented, diluted and marginalized, and thus became politically acceptable. The Green movements, for instance, evolved into Green parties that gradually accepted numerous compromises, became parliamentary parties and blended into mainstream politics. This was now part of an overall, impotent and non-threatening 'diversity'.

The reaction to the very possibility of a serious challenge to the world order was so profound that today it has become 'quite difficult to imagine a time when people once envisioned a world different *in essential ways* from the one in which we now live' (Ross 2002: 20; my italics). We needed to wait for the situation to worsen, probably beyond repair, and for an American celebrity politician to get the words 'environment' and 'crisis' back onto the decision makers' agendas. Al Gore's *An Inconvenient Truth* (2006) has had an enormous media impact. It created a surge in interest in sustainability and helped to raise a strong collective awareness about environmental problems. Today, even the most conservative politicians cannot avoid acknowledging the grim reality. The ecological crisis is being addressed through numerous research projects and solutions are being sought – from the broadest global scales down to the smallest of local actions.

But it is already obvious that the simple act of bringing sustainability into the media focus is not going to produce significant results. The overall value system remains unaffected. Many high-quality projects are being overshadowed by even more initiatives and undertakings that operate with a mindset that does not reach beyond the fast dollar. The seemingly unstoppable biofuel fiasco in the USA may be a good example of how even a good intention, when under-researched, hastily implemented and propelled by greed, can produce terrible detrimental outcomes (*Time Magazine* 2008).

What matters to us in the context of this book is that the existence of a global environmental crisis is now acknowledged, that *environmentally* sustainable development is being seen as an urgent need, and that it is at the top of the decision makers' agendas. We stress the urgency of a radical shift towards ecologically responsible urban development, and also put forward a reminder that it is critical to remember the other side of the dialectical couple which constitutes acceptable development – *cultural sustainability*. Over the last couple of centuries, and particularly over the last decades, cultural environments and social diversity were under threats which were as strong as those that threaten and destroy the ecosphere and environmental diversity. Sociocultural richness all around the world is vanishing as rapidly as the other endangered species are. Both the ecosphere and the socio-sphere are endangered by the same forces, which are often represented as one of many expressions of globalization.

Our critique of the sorry state of the global political, sociocultural and environmental reality sketched above focuses on the production of space. The production of space is framed by the key concepts of globalization, sustainability and urbanity.

GLOBALIZATION

It is impossible to mark globalization with a denominator that has a single value. The term itself is, indeed, one of those that get 'used with more frequency that understanding' (Whitehead 2007: 118). Simultaneously a process, a condition and a political project (Harvey, in Scholar 2006), globalization embodies the potential for some of the most fascinating and some of the ugliest expressions of humankind. It encompasses both an enlightening new capability to communicate and to share globally, and those old colonialist instincts towards domination.

A wide range of sociological and cultural theories chart the maze of possible and probable definitions of globalization. Terms such as 'overlapping discourses', 'communications revolution', 'time–space distancing', 'time–space compression', 'global consciousness', 'sociology beyond societies', 'borderless world' (as summarized in Ray 2007: 7) map the field of possibilities. However, a closer look at actual practice in almost any field of human endeavour which is conducted on the basis of those increased flows, extended reach, new overlaps and spatio-temporal compressions exposes a picture of an increasingly unjust and polarized world. While many see globalization as the latest stage in a progressive evolution of humankind, others experience it as an exploitative process of steady decline, a descent into an unbearable homogeneity.

Homogeneity, again, is seen by some as an ideal basis for limitless communication, while at the same time many fear its reductionist force, its acceptance of the lowest common denominator, its reduction of social complexities to the simplicity of the free market, where citizens become consumers and thus, in a cynical turn of phrase, 'citizens of the world'. A huge body of literature glorifies the free flow of capital, goods and people from which we should all benefit, while an equally rich and numerous volume of works exposes globalization as a chaotic and destructive form of neo-colonialism, as pure ideology. 'In some ways', Larry Ray argues, 'the old territorially based ideological divide between capitalism and communism has been replaced by fluid transnational identity politics and anti-globalization protest movements.'

From Harvey's Lefebvrian tripartite definition of globalization as process, condition and political project (Whitehead 2007: 119) emerges an understanding of globalization as the *geographical reorganization of capitalism*: in other words, the old story, rewritten large, with recognizable patterns of re-energized colonialism and the continuity of established power-structures, and their combined unwillingness and incapacity to change their course. That picture, combined with the necessity of radical change, does not leave much space for optimism.

It seems that the very term *globalization* has become compromised to the extent that we may need new words and new concepts that are open enough to facilitate the emergence of new agendas. Ortiz thus suggests the parallel use of terms *globalization* and *mondialization* in what is not just a provocative word game. He argues that, while the term 'globalization' still may be applied to the economic and technological spheres, 'mondialization' is better adapted to the cultural universe (Ortiz 2006: 401). This proposal has broad connotations that emerge from the implied departure from a single denomination to parallel definitions, from the use of a single language to the recognition of the existence and importance of (many) parallel languages (*globus, globe, mundus, monde, mondo* – all the same and different at the same time).

'Mundialised culture', Ortiz continues:

> promotes a cultural pattern without imposing the uniformity of all; it disseminates a *pattern* bound to the development of world modernity itself. Its width certainly involves other cultural manifestations, but it is important to emphasise that it is specific, founding a new way of 'being-in-the-world' and establishing new values and legitimizations. And that is the reason why *there is not and there will not be a single global culture, identical in all places. The globalized world implies a plurality of world-views.*... And this means that *globalisation/mundialisation is one and diverse at the same time.*
>
> (Ortiz 2006: 403; my emphasis)

Within such a doubly coded process, environmental sustainability, which implies environmental sensibility, responsiveness and responsibility, should have a decisive role. Environmental sustainability, as proposed above, should always be inseparably coupled with cultural sustainability in a dynamic, dialectical pair that frames the thinking about, and the realities of, *the urban*.

Focusing on cities, this book deliberately merges ecological/environmental and cultural/urban sustainability, and recognizes their (ideo)logical flow into – *eco-urbanity*.

ECO-URBANITY

Two well-established, but vaguely defined, concepts make *eco-urbanity* – *ecological* sustainability, and sustainability of *urban* cultures. When defining sustainability, it is unavoidable to refer to the Brundtland Commission's report in 1987. This famously stated that 'sustainable development is a development that meets the needs of the present, without compromising the ability of future generations to meet their own needs'. Over the decades, that definition has attracted a lot of discussion, and this in itself deserves a certain merit. Such a broad and all-encompassing definition attracted equally varied critiques, counter-proposals and developments, and sought various contextualizations. Some of those evolved into 'sectoral' definitions. In order to give a general sense of directions taken, I table only three (almost), randomly selected definitions. *Ecological* sustainability is, for instance, seen as 'the ability of a society, ecosystem, or any such ongoing system to continue functioning into the indefinite future without being forced into decline through the exhaustion or overloading of key resources on which that system depends' (Global Sustainability, n.d.). *Cultural* sustainability is often defined as 'improving the quality of human life while living within the carrying capacity of supporting eco-systems' (WWF 2008), while some see *economic* sustainability, in the context of overall sustainable development, as a promise that 'economic activities can be coordinated with environmental protection and that technologies can be implemented to ensure that economic growth does not damage the environment and can be solved without social and political disturbance' (Beders 1993: xi). Even those few definitions show both the need for and the trappings behind defining sustainability. To put it simply, the issue is laden with values, and definitions can and do come with all ideological twists imaginable. The most interesting example of an ideological struggle in the definitions above is Beder's notion that sustainable development has to make sure that the shift towards sustainable development (safely reduced to 'environmental *protection*' and technology) can be solved without social and political disturbance.

Such narrow views cannot help. Good definitions of sustainability tend to be composite and complex, and they always remain incomplete. I find them at their best when they generate fields of forces that are only represented and captured by words, and that can and should frame concrete actions and give concrete results. The common question of how to implement such broad definitions can be answered only through action. The value basis of true sustainability is a philosophy of action and its ultimate test is always in its local relevance. Definitions need to seek a local anchorage, circumstantial adjustments and contextualization. In the case of the production of the built environment, this means that appropriate definitions can exist only in the function of appropriate planning, design and construction practices and products.

In their *Blueprint for Sustainable Australia*, Krockenberger and Thorman (2000) aim to be all encompassing. They write:

> sustainability is a dynamic process that enables all people to realize their creative potential and improve their quality of life in ways that simultaneously protect and enhance the Earth's life-support systems and its variety of life. Sustainability is not the pursuit of a steady state, but a dynamic process. It is for all people, of this generation and those following. And it has a primary goal of ensuring fresh air, clean water, healthy soil and protection of nature. Fairness is embodied in the concept of sustainability – fairness to this generation, to following generations and to other species.
>
> (Krockenberger and Thorman 2000)

This definition weaves together a dense tissue of concepts that include dynamism, empowerment, creative potential, quality of life, protecting and enhancing the Earth's life-support systems, variety of life, all people, all generations, fresh air, clean water, healthy soil, protection of nature, fairness. That composite field can act as a springboard for development of ideas, the quality of which needs to be tested through action, which recognizes that the Latin term *sustinere* is an adjective and is critically dependent on concretization.

It is also worth noting that Krockenberger and Thorman's tone is self-consciously emotional. They declare that the quality sought was 'for all people, of this generation and those following', that the fairness sought is 'fairness to this generation, to following generations and to other species'. This statement challenges the common positioning of sustainable solutions as the exclusive protectorate of science and engineering. When thinking about sustainability we must acknowledge that, as Low has argued in another context, 'those processes are not simply complex in the sense that they are technically difficult to grasp (though this is certainly often the case). Rather, they are also complex because they *necessarily exceed our capacity to know them*' (Low 2004: 6). Dealing with sustainability has to include both measurable and non-measurable, quantifiable and non-quantifiable, and the understandable aspects of the environment and our engagement with it, some of which may escape reasoning.

Environmental sustainability and social sustainability, located in an urban node, make *eco-urbanity*. The key terms that constitute the elusive realities of sustainability and urbanity are dialectical. Their interplay implies the blurring of artificial boundaries between globalization and localism, the ecological and the cultural, the urban and the architectural. The only way towards truly sustainable development is integrative, which does not separate the environmental and the cultural, the natural and the human, and which transcends binary oppositions and sees opposites in a dialectical rather than opposing relation.

The concept of urbanity is no less elusive than the concept of sustainability. In a single word it captures the essence of *the urban*, pointing at what makes cities – cities. And, according to Michel de Certeau – the city is that 'most immoderate of human texts' (Highmore 2005: xi). The concept of urbanity is usually associated with, and is usually translated, as *citiness*. But, when we think across the diverse cultures that are considered *urban*, what do we really mean? What are the real meanings behind, and the expectations from, say, the Thai concept *kwam pen muang*, the Serbo-Croatian *gradskost*, the Italian *urbanita*, and the Japanese *toshisei* – all of which claim to be translations of the term 'urbanity' (Radović 2005b)? Such meanings and expectations are, and they have the right to be, as diverse as Thai, Yugoslav, Italian and Japanese cultures in their totality. Urbanity is, thus, just a term that is increasingly accepted and which, by referring to something so culture-specific, resists translation.

As the concept of urbanity is so qualitative, defining it invites subjectivity, for our subjectivities are equally culturally framed. For instance, *I like* the French saying *L'urbanite – savoir faire la ville, savoir vivre la ville*. 'Urbanity – knowing to make the city, knowing to live the city.' (But is *savoir* really equivalent to *knowing*?) That saying is, probably, one definition that may be comfortably applied across the cultural diversity of human universe because it is loose enough.

Urbanity is commonly used as a synonym for urban culture (Zijderveld 1998). 'Urbanity is not a static situation, ideology or state of mind. It is, as culture and language, the process of consolidation of behavioural codes in a community' (Bobic 2004: 39). That implies the critical importance of locality, of concrete place, concrete social circumstances and concrete culture. The original term comes from *Urbs*, an old, formal name for Rome. In its original form, it referred to a particular kind and level of refinement, to manners that were to be expected only from *homo urbanus*, the select group of inhabitants of the very centre of the empire (Ramage 1973). Interestingly, as the Roman world spread (in an early example of globalization by military might), the term 'urbanity' also started to stretch. It begun to lose its exclusivity and ended encompassing all citizens of the empire.

A number of parallels can be drawn between that process and current globalization. Today, global financial forces (often still supported by military might) also tend to question and flatten difference. They introduce their own imprints onto the ground over local patterns, most commonly regardless of local environmental and cultural conditions. The global capital with its World City fetish (Radović 2008), for instance, seeks recognizable similarities in the cities it wants to impregnate. Such World Cities are certainly not of the quality implied in Johann Wolfgang von Goethe's original notion of *Weltstadt*. Goethe 'was using the concept to identify the leading *cultural* centres of the world' (Taylor 2004: 21; my italics), as opposed to the leading centres of financial might of today. A Goethean *Weltstadt* would cherish centres of *urbanity* that are understood and fully lived locally, as *kwam pen muang*, *gradskost*, *urbanita* or *toshisei*. This quality is rapidly disappearing under the imperatives of a rampant, 'free market'-driven globalization. In order to reclaim the appropriate cultural position of *the urban*, and thus regenerate its creative energies, we need to reclaim socially sustainable development. That implies true mondialization and pointed questioning of dominant 'econocentric' thinking. Thinking based on the idea that everything should 'be solved without social and political disturbance', is dangerously out of date (Saul 2005). Solutions are becoming impossible. Both social and ecological sustainability cry for a paradigm shift. The new condition demands a new economy, critically marked by sociocultural responsiveness and responsibility (as if people mattered, Schumacher 1993) and, certainly, not by greed.

I started using the term 'eco-urbanity' (always with a small 'e') in an effort to stress that the concepts of ecological sustainability and urban culture belong to the same value system, one that calls for ecologically sustainable development while simultaneously celebrating urban culture. In simplest terms, urbanity always was and will always remain concerned with behaviour. It concerns being *urbane*, expressing certain manners in relation to neighbours and fellow citizens. Urbanity is *socially* responsive and *socially* responsible behaviour. If we extend such considered, well-mannered behaviour towards whole regions, towards the broadest environments we inhabit, we create *environmentally* responsible and *environmentally* responsive behaviour. In the first case, we are good citizens of our social milieu; in the second, we are good citizens of the world.

Urbanity and sustainability are, thus, far from hostile to each other. They are two aspects of a single sensibility that departs from the narrow anthropocentric views that shape our current way of dealing with the environment and brings us closer to a broader, eco-centric position. Eco-urbanity is a sensibility that brings together responsiveness and the responsibilities necessary for an environmentally and culturally sustainable future. Environmental and cultural sustainability cannot be separated (Radović 2005a). As action-oriented values, they stem from a strongly profiled, dynamic value base and are in constant flux. Sustainability and urbanity can never be taken for granted. They need to be nurtured and perpetually recontextualized.

eco-urbanity is an ability to create and to live in environmentally and culturally sustainable cities.

Towards the production of sustainable urban space

Redefining urbanity to fit contemporary imperatives of ecological sustainability demands new ways of thinking about *the urban*, which embraces an environmental imperative. Lefebvre's powerful demand (1996) for *le droit à la ville*, the renewed right to the city, and for *le droit à la différence*, the right to difference, now include a right to a sustainable urban future. Rebelling against the homogenization, fragmentation and hierarchically organized power of globalization, they demand an inclusive, mondialist local sensibility.

There is nothing idyllic about sustainability and urbanity per se. Sustainable conditions may mean profound suffering and an anti-democratic, if not an utterly anti-human, social order (Ferry 1995). Urbanization has always included tensions, oppositions, controversy and conflicts. Neither of those concepts is thus intrinsically positive. Together, they only provide the frame of a promise of a better future. The imperatives of our environmental and cultural crisis demand from us the sensitivity to bring together these two concepts into a dialectical movement towards an ideal that we have to strive for. There are many possible paths towards an uncertain goal. Those paths create an open field, and navigating it is not easy. The journey towards the sustainable city demands a compass in the form of a solid set of guiding values, which themselves remain open to rethinking and adaptation.

The journey demands both creativity and innovation, and rootedness. Creativity and innovation are needed to make sure that at every moment each actual condition on the ground is as sustainable as possible. A self-confident identity is a necessary precondition for cultural survival. Urban design, landscape design and architecture for a sustainable future need to operate in that dynamic field. They generate physicality, spatial expressions of the guiding values that reflect the range of their loco-temporal variations. And as there is no end to that journey, there is neither stasis, nor some definite, achievable ideal sustainable city. As environmental conditions continue to change, the process of perpetual reinvention must continue. That is what makes cultural sustainability, the stability of own point of departure, so important. More sustainable futures will depend on our inventing ever-new urbanities, and those are possible only if they, at the same time, keep on confirming local tradition. The 'new', sustainable city needs to help us constantly rediscovering 'colour, noise, gradients, materials, the rhythm of lighting, the atmosphere of the city full of accidents. Surprise must remain possible, the eruption of the non-assigned, the force of diversion, sudden and then gradually accepted' (Barré 1980: 7). At every moment, the sustainable city has to be a just, fully lived city. It has to be strongly informed by the dialectical relationship between the global and local, and the environmental and cultural responsibility of its citizens. Each city itself has to be a responsible collective citizen of the world, while remaining accountable to each and every one of its own citizens. A sustainable city has to be confident about its own past, exist in the present time, and project a multiplicity of trajectories towards possible, desirable futures. Those trajectories have to be based on a strong value system, eco-urbanity.

In the world of an increased eco-cultural responsibility, locally responsive environments do not deny the importance of universalizing forces. On the contrary, as the projects presented in this book amply confirm, it is exactly by confirming their local identities that cities and architecture reach true global quality. The environmentally and culturally responsive and responsible spaces, buildings and activities that make such cities are spatial expressions of eco-urbanity.

BIBLIOGRAPHY

Barré, F. (1980) 'The desire for urbanity, Paris Biennale', *Architectural Design* 11/12.
Beders, S. (1993) *The Nature of Sustainable Development*, Newham: Scribe Publications.
Bobic, M. (2004) *Between the Edges*, Bussum: THOTH.
Bookchin, M. (1980) *Towards an Ecological Society*, Montreal and Buffalo: Black Rose Books.
Bourg, J. (2007) *From Revolution to Ethics*, Montreal and London: McGill–Queen's University Press.
Brundtland Commission (1987) *Our Common Future*, Oxford: Oxford University Press.
Ferry, L. (1995) *New Ecological Order*, Chicago, IL: Chicago University Press.
Fromm, E. (1976) *To Have or To Be*, New York: Harper & Row.
Global Sustainability Online (n.d.) Available online: http://globalsustainability.org/education/definitions (accessed May 2008).
Gore, A. (2006) *An Inconvenient Truth*, Emmaus, PA: Rodale Press.
Highmore, B. (2005) *Cityscapes: Cultural Readings in the Material and Symbolic City*, Houndmills and New York: Palgrave Macmillan.
Krockenberger, K. and Thorman, R. (2000) *Natural Advantage: A Blueprint for Sustainable Australia*, Melbourne: Australian Conservation Foundation.
Lefebvre, H. (1996) *Writings on Cities*, Oxford and Cambridge, MA: Blackwell.
Lovelock, J. (1979) *Gaia: A New Look at Life on Earth*, Oxford and New York: Oxford University Press.
Meadows, D.H., Meadows, D.L., Randers, J. and Behrens, W.W., III (1972) *The Limits to Growth*, New York: Universe Books.
Ortiz, R. (2006) 'Mundialization/globalization', *Theory, Culture & Society* 23, 2–3: 401–3.
Radović, D. (2005a) 'Think about the soul of Bangkok! – fragments from an investigation into *kwam pen muang, gradskost, urbanita, toshisei*', *Raeang* 4: 99–107.
Radović, D. (2005b) 'Urbanity and sustainability', in *Sustainable Buildings 05 Conference Proceedings*, Tokyo.
Radović, D. (2008) 'The world city hypothesis revisited: export and import of urbanity is a dangerous business', in Jenks, M., Kozak, D., Takkanon, P. (eds) *World Cities and Urban Form*, London and New York: Routledge.
Ramage, E.S. (1973) *Urbanitas, Ancient Sophistication and Refinement*, Norman, OK: University of Oklahoma.

Ray, L. (2007) *Globalization and Everyday Life*, London and New York: Routledge.
Ross, K. (2002) *May '68 and its Afterlives*, Chicago, IL: Universty of Chicago Press.
Saul, J.R. (2005) *The Collapse of Globalism and the Reinvention of the World*, London: Penguin, Viking.
Schmid, C. (2006) 'Theory', in Diener, R., Herzog, J., Meili, M., de Meuron, P. and Schmid, C. (eds) *Switzerland: An Urban Portrait, Introduction*, Berlin: Birkhauser: 163–224.
Scholar, R. (ed.) (2006) *Divided Cities: the Oxford Amnesty Lectures 2003*, Oxford and New York: Oxford University Press.
Schumacher, E.F. (1993 [1973]) *Small Is Beautiful*, New York: Harper & Row.
Taylor, P.J. (2004) *World City Network: A Global Urban Analysis*, London and New York: Routledge.
Time Magazine (2008) 7 February.
World Commission on Environment and Development (1987) *Our Common Future*, Oxford and New York: Oxford University Press.
WWF (2008) Available online: http://abc.net.au/architecture/ar_mur.htm (accessed May 2008).
Whitehead, M. (2007) *Spaces of Sustainability*, London and New York: Routledge.
Zijderveld, A. (1998) *A Theory of Urbanity: The Economic and Civic Culture of Cities*, New Brunswick, NJ and London: Transaction Publishers.

2 THE BARCELONA AGENDA
Reuse, compactness and green

ORIOL CLOS

INTRODUCTION

For many centuries Barcelona has been growing upon itself within strictly defined geographic limits – the Serra de Collserolla mountains, the Besos and Llobregat rivers and the Mediterranean Sea. As a result the urban development of Barcelona illustrates a desire to save space through careful and strategic town planning. Over the years Barcelona has developed a variety of specific spatial and development strategies. Here I want to highlight three fundamental lessons that Barcelona has learned from its own past and which continue to shape the spatial organization of the city.

The first lesson is to redistribute land use over the city. This should not be understood as a continual expansion of the urban fabric and occupation of

Figure 2.1 Barcelona.

new spaces. Recovery and rethinking spaces that have lost their original use is an important task, both in visions for the inner metropolis and those that deal with the 100 km^2 that encompass the municipality of Barcelona. The transformation of the old factories is the way to meet current needs and provides the basis for the overall urban reformulation of Barcelona.

The second is to define the levels of urban intensity in the most central areas. An ongoing task is to rethink the basic urban parameters: density, use and land occupancy. New ways to define these reference indexes provide more complex formulations of the base concept and incorporate qualitative evaluations of urban spaces and their environmental characteristics. Using a combination of indicators of intensity, distribution and mixed use are the key parameters of designing the compact city.

The last is to recognize that large open spaces are part of the city, a component that balances the ratio between areas of great intensity and open zones. This new evaluation is based on acknowledging that different values are associated with different kinds of spaces, their material features, their internal and external qualities and their latent development potential. They are essential elements in the city and should not be seen as vacant land waiting to be purchased and developed, or be subjected to great pressure, both public and private, to be modified.

TRANSFORMATIONS IN THE PAST

Over the last 25 years Barcelona has developed a number of fine examples of giving its land new purposes for a number of new city projects. The difficulty in finding free land within the city limits and its surroundings has been conducive to inventing creative new uses for many once-obsolete and derelict inner urban areas. Those spaces, originally envisaged as large, segregated precincts for military, railway, sanitary, infrastructural and manufacturing purposes, now accommodate new poles of the values of centrality. Many still wait to be integrated into the city.

Existing urban areas are being transformed in a number of ambitious urban projects. These action plans have their own physical, temporal and economic limits. The programmes developed in these plans are complex and have been given strong public support, allowing the development of new landmarks and reinforcing the value of these projects in an established urban landscape. These new initiatives redefine the use of parcels of lands that are important to the city, both in terms of their location and development potential, which is often far greater than their current use.

Starting from the second half of the twentieth century, many general plans of action, known as urban projects, have been implemented throughout Europe. Complex projects of this type have established the basis for the plans and actions developed in Barcelona after 1975. They marked a thorough transformation of the city, culminating in the Olympic Games in 1992. In the first pre-Olympic phase, the key aim behind the plans was to transform the city by renewing its key public spaces. The public sector was the most important driver of this process and, with the power embodied in the city authorities, successfully promoted this transformation. In the next period the focus shifted to the urban fabric. A more decentralized administration, with the agreement with the private sector, started to agitate for this change. At this stage, nearing the Olympic Games that took place in 1992, the strategy was to recover large areas of land and to open connections between the city and the sea. The focus was on large developments along the coast and towards the east, and in increasing the value of central areas by the construction of an efficient ring road.

Olympic Barcelona redirected the evolution of the city. The year 1992 defines the starting point of a number of innovations that are still being developed today. The structural options have been consolidated by territory from the inner ring of the cluster, slightly greater to the administrative surroundings of Barcelona, defined by clear geographical boundaries as lying between the sea, the mountains and two small deltas. Using these physical reference points, what is strictly considered to be the city of Barcelona lies between the two rivers. The adjacent municipalities, with their related morphological continuity that are like fragments of the compact city, spread in territorial scope from the natural areas within the city to the mountains and the sea. It also encompasses the great infrastructural plans of the countryside, the port and the airport, and their logistical facilities. Certain aspects of urban development policy and the redefined socio-economic planning of the city were established around the outer proximities of the municipality. Still pending, like a structural deficiency in Barcelona's development, are unresolved political and administrative issues: revitalization of the mode of metropolitan government and the organization of territorial strategic urban planning.

The city that was defined and constructed in 1992 constitutes the physical base for the new qualitative leap Barcelona is preparing to take. The city must now get to grips with new conditions and challenges that include the distribution of production, the growth of immigration, the need for accessible housing, service demands, diversification of the means of transport, environmental requirements and opposing social, economic and cultural interests. In this context, city planning must adjust to the changing conditions that take place within the city and regulate the processes of urban development, overcoming the exclusive arrangement of permanent physical conditions. It needs instead to be strategic, not just using projects merely as tools for urban planning but in the understanding that processes are more important than projects and we need tools to manage them. Around the year 2000, a turning point in Barcelona's urban planning was achieved by incorporating new resources to face new challenges and by redefining certain strategic objectives.

TRANSFORMATIONS IN THE PRESENT

In Barcelona today, urban projects are no longer defined as ways of substituting for obsolete uses, but are conduits of more complex processes of transformation. This requires replacing the precision of previous projects to resolve clear and specific issues with open processes that are able to adapt to uncertainty, in order to accommodate long-term transformation and be more responsive to a dynamic reality. By incorporating this new way of defining the rules of urban evolution, different situations can be better resolved than before. This is our way of responding to four aspects of development which were recognized to be fundamental to the successful urbanism of Barcelona in the coming years:

- *Urban fabric*. Programmes must express the complexity of the contemporary city, where new customs and landscapes mix. That is, an urban design of mixed criteria is needed when it comes to distributing land use on multiple scales.
- *Public space*. The city is a place that brings people together and unifies the diverse aspects of what goes to make up the city. Public space is the infrastructure that supports urban development and is thus a primary planning instrument and underlies the control of both public and private development.
- *Intensity*. This concept expresses the increasing complexity of cities as locations for large events that enrich social development and is an instrument for defining quantitative values in order to customize density, mix of uses and the distribution of a variety of spatial categories.
- *Empty space*. Defined, within the parameters of a compact city, by contrast with what is taken up, such spaces include irregularly distributed areas on a grand scale. The key idea is to integrate what has not been occupied and make it an inherent part of the overall urban body.

Within the framework set by these broad objectives, the contents of the plan are managed by imagining pieces of a city in the near future, taking into consideration many social sectors that increasingly interact in their urban surroundings. Barcelona has given priority to massive housing construction within the framework of regional and central governmental policies that adhere to different plans and projects. The search to provide a richer and more varied urban fabric has also led to redefining the regulations on productive activities so they can be integrated in mixed spaces in the urban centre.

These two components of urban fabric allow some city neighbourhoods to bring themselves up to date, and others to absorb new neighbourhoods in areas of lands that were previously separate from them and have now been recovered from past use. In such conditions, green zones are fitted out and established as scaled complex systems that unfold over an area. The best expression of urban balance is found in heavily populated areas with a cohesive core of small public spaces that have a local feel and nearby facilities, understood as a city within a city. Mobility, and its public manifestation, public transport, is the structure that brings the citizens together and guarantees universal accessibility to all spots in the city.

The plans and projects that are currently being developed in Barcelona fall into the following categories, based on established planning methods and the strategic role of urban transformation:

- *The process*. Reusing the land permits the development of a coherent new fabric with great strategic objectives: building housing and workplace structures adjacent to each other. The transformation plans set up organizational guidelines for far-reaching processes to configure new morphologies and urban structures. All this is done by acknowledging pre-existing physical structures and their improvements, by supporting new developments, streets, and pre-existing land divisions and buildings, and by seeking a seamless physical and practical transition between the original and the future state. Renewing the infrastructure that incorporates public space is the way to gain public support for new projects, allowing a new fabric to be developed over the years and adapting concrete solutions to the need for urban evolution.

- *The structure*. The public space, the transportation infrastructure and large facilities underpin the establishment of new urban structural relations. To overcome progressive urban fragmentation and create continuity between disconnected parts, the centre needs to be consolidated and strengthened. The new structural works reinforce a system of axes and infrastructure that offer morphological resources to the entire city and that integrate it into a singular urban concept of buildings, infrastructure, free space and services.
- *Empty*. Large free areas play a balancing role in the transformation of a compact city. They are read and treated as undeveloped spaces in the urban structure and locate the morphological and historical laws that have actively participated in defining the surrounding urban fabric. A set of criteria is now needed in order to amalgamate these areas into new environmental, recreational and educational spaces for the citizens.
- *Enclave*. Urban projects are planned in very specific surroundings of varied sizes and under different programmes. They are based on architectural plans that were themselves responses to previous demands, or that respond to a new opportunity arising from the availability of some newly obsolete space. The programmes cover a wide thematic range, from needs for individual functions to situations of great complexity where different uses and urban elements are combined to provide a richer texture of use.
- *Reform*. Plans for facilities and public spaces in old neighbourhoods and structures complement the actions for strictly social and economic new and improved housing. Modifications are needed to enact current plans that were drawn up more than 20 years ago, in order to overcome the financial shortcomings of some of the more modest neighbourhoods of the city. The establishment of a process of taking small actions for renewal continues to be a resource in such situations by knitting together the unfolding housing projects, the proximity of facilities and public space.

All these plans, studies and projects make up part of a global, coherent and united strategy to transform the entire city. The specificity of each action extends both quantitative and qualitative urban value to the whole city. These urban strategies unfold in an unordered, non-hierarchical manner. Beyond the scheduled plans, opportunities of all types mark the development of action plans, establishing a heterogeneous totality of solutions, backed by wide-reaching guidelines. Each plan and each project applies its own singular and integrated force to show that the whole is richer and more complete than the sum of its parts.

PRACTICAL EXAMPLES OF URBAN RENEWAL

Some of the plans being developed in Barcelona over the last few years, framed at different levels of complexity and specificity, illustrate precise and partial aspects of general arguments for urbanism today. They are practical examples for research and debate that can be used in other contexts to argue about the sustainable city. Besides the need to establish a set of tools for precise objectives and to solve problems, there are real-life urban surroundings that provide the opportunity to test new plans of action. In Barcelona, three research-based plans support my specific arguments:

- The urban-renewal process is based on understanding that urban ways persist. This gave us a programmatic basis for plans for the city that existed previously in the 22@ plan, and allowed us to refocus on the duality of permanence versus substitution.
- The urban parameters on the new Marina del Prat Vermell neighbourhood allowed us to study and make the case for density indicators and mixed use in the compact city.
- The studies defining the strategic green aiming at large open spaces in the city allowed us to reinforce the definition of compactness as the central urban paradigm for Barcelona, as a location balanced between population concentration and free space.

Reuse, compactness and green space come together in defining an urban reality based on saving land and on concentration, to optimize the efficiency of services and ensure an intensity of use and urban experience. In this frame of reference, the need for a balanced land distribution is established both locally and regionally. These principles of reuse, concentration and balance form part of the body of Barcelona's history and are currently being updated and made ready for the future of these essential elements of Mediterranean urbanism.

The model of the compact city, using the transformation of Barcelona as a guide, is achieved by modernizing some set arguments. Distance and density, as expressions of physical proximity, mixed distribution of uses, the cohesion of core urban values in housing and the proper use of infrastructure and open space, as a universal strategy of scale and service, are the topics that define compactness as a goal for city development.

Persistence of the urban structure in the 22@ plan

In the year 2000, Plan 22@ was approved for the old industrial sector of Poblenou, which for a century-and-a-half contained the greatest concentration of factories in Catalonia. The plan was a strategic challenge to maintain the productive capacity of the city and encourage the introduction of economic activities compatible with residential development and the efficient use

Figure 2.2 The urban structure in the 22@ plan.

of space. It aims to steer a progressive transformation of more than 200 ha of the existing city. This process is to be developed over the course of many years, supported by Barcelona's network of streets and updating urbanization the supporting infrastructure and services that run through these streets.

Plan 22@ regulates the way in which the existing fabric will be transformed by progressive increments of use and density. The planning tools are flexible, in order to adapt every urban action to the conditions on the ground in each location and to suit the temporary steps required for seamless physical and social implementation. Incorporated in the plan's objectives and standards of development is a balanced combination of permanent elements together with the substitution of existing elements. It is therefore designed to tackle the urban and cultural challenges of a persisting urban structure.

There are a number of permanent elements that the transformation of Poblenou must be built around. I mention here renewing and upgrading the streets laid out in Cerdà's 1859 extension to support the current city development. These 20 m-wide streets, forming a grid of 133 × 133 m, are progressively renewed to facilitate efficient city services in a revitalized industrial district. By maintaining this infrastructure, the city guarantees a continuation of urban activity while piecemeal developments are being made to transform the old industrial sector. The public management and control of the process is based on a plan that establishes five basic aspects of these streets: energy, communication, water, waste and mobility.

Existing housing is to be integrated in the new mixed fabric in order to maintain the continuity of the façades that characterize the urban space and to guarantee the social continuity of the resident population. The housing that is retained, together with the new public housing inserted in the neighbourhood, provides both residential activity and urbanity. Poblenou has the fortune to be situated in a central business district that is itself 50 per cent residential.

The Cerdà block is defined as the minimum size of intervention. It covers approximately 12,500 m^2, comfortably permitting space for new buildings and providing for a combination of distinct volumes, thus conserving pre-existing elements and avoiding a blank-slate type of new development. In all urban proposals public facilities and service spaces are incorporated. There are two kinds of existing factories that need to be considered. There are buildings with historical and national significance, depending upon their characteristics or their ability to evoke the productive and industrial past of the neighbourhood, which should be preserved. There are also more recent industrial buildings that are normally large, yet very compact, that may be able to be converted to use by the technological sector, assuring their continual productivity.

These options for maintaining and reusing existing elements are conducive to a seamless process of urban evolution and to adapting the land to new uses. This offers a way to define the basic zoning patterns by acknowledging the different ways in which land has been occupied in Barcelona's urban expansion. Flexible zoning solutions allow the establishment of urban structures with formal and symbolic references to those already existing. Permanence and substitution are dialectical attitudes for expressing complexity and they can be used to manage the accidents of urban planning, at times in contradictory ways, but without falling into the trap of developing, on the one hand, a historical impoverished space or, on the other, a formal, picturesque, yet fragmented, style.

Intensity in the La Marina del Prat Vermell plan

As a response to the strategic lines of concentration and optimization of the resources being considered in many locations, the conditions for the orderly densification of land need to be established. To do so, the plans and the regulations that they cover have to be rethought and readapted, reassessing the indexes of building rights, density, compactness, capacity, distribution of use and relation to the system of public services.

To achieve dense compactness and recentralize the city, a multi-purpose urban fabric needs to be created to combine diverse yet compatible urban use in the framework of socio-economic heterogeneity and the goal of optimizing land use. Housing, especially public housing, is what will bind the new urban fabric. The new productive activities that will be part of this mixed urban fabric are compatible with housing and allow for mixed use in the same environment, even in the same building. The current organization of economic activities needs the urban norms for the zoning and specialized planning of land use to be modified. New tools for regulating compatible use and coexistence need to be defined to include a range of dynamic activities in the economy and high value-added industry.

Barcelona's City Council has approved the 2004–10 Housing Plan to improve access to housing for the people of Barcelona. Under this plan the residential needs of the city have been assessed. As a result, 30,000 new dwellings will be constructed in urban transformation areas, reusing old industrial land. Housing will be constructed in large groups so as to structure

Figure 2.3 La Marina del Prat Vermell plan (Muro-Lay, 2007).

new neighbourhoods and rebalance the distribution of land use throughout the city. These neighbourhoods will be seamlessly integrated into the urban scheme, with relationships of continuity and connectivity with the existing fabric. Each housing group has its own character based on the morphology of the new urban fabric, and in relation to their own internal structures and the landscapes and landmarks that shape them.

La Marina del Prat Vermell is the first area to be developed following the pattern laid out by the housing plan. It covers a total of 72 ha of industrial land of low-intensity use, situated in a sector of Barcelona tied to the important transformations for taking in large facilities and city services. The construction of two new metro lines will complement the other new infrastructural developments and radically improve the accessibility of these locations, fully integrating them into the central urban system.

To manage this transformation of formerly industrial areas, planning criteria are needed to establish density benchmarks for the new housing districts. All around Europe, Mediterranean cities are being used as models for density in recognition of the quality of urbanity, services and greater levels of social cohesion in their denser areas. The general term 'density' relates to numerical ratios for land use, compactness, gross and net buildable area and number of houses. We name this concept 'intensity'.

As a model, the distribution of land use for building is 40 per cent, leaving 60 per cent for roads, open spaces and utilities, resulting in a compactness ratio of 50 per cent (10 per cent utilities), which allows large, open spaces to be left within the fabric of the town. The gross buildable area index is $1.6 \, m^2/1 \, m^2$ which, in these proportions of land distribution, gives a net buildable area index of $4.0 \, m^2/1 \, m^2$. For mixed, residential–economic activity land use with a proportion of 70 per cent to 30 per cent, the global density of housing is between 110 and 140 homes/ha. This model for applying buildable

ORIOL CLOS

area indices to land distribution parameters with the emphasis on public systems has already been tried out in several cases under study, allowing us to balance the need for open space and facilities and define a fabric that is capable of solving the requirements of the new population that will inhabit them.

The strategic green

In the idea to balance the compact city with intensive-use areas and free open spaces, we developed the idea of 'strategic green' as an area with a structural role in the configuration of the city. Our analytical and definition studies class large open spaces in the city as an inextricable part of urban space, either already consolidated or in the process of transformation. Collserola, els Tres Turons, Montjuïc and Ciutadella are areas that have been studied within the frame of this strategy. They have the capacity to contribute greatly to urban change as well as to promote equilibrium in the distribution of ground space

Figure 2.4 Balance in the strategic green plan.

in the city of Barcelona. The main idea is to understand these large open areas as active parts of the city, not just left-over space that is still to be developed. The renewal of public areas and the recovery of environmental values, particularly those associated with green areas and private mobility, are integrated completely into the urban structure, and they lead to the establishment of new relationships between heterogeneous parts of the city where open space is a valid part of the urban continuum.

The logic of having empty spaces interspersed with densely populated surrounding areas provides a new way of defining the role of these spaces in the framework of the city. The plan is not to infiltrate green areas into urban space, nor mix, nor occupy open ground; it is to establish relationships between areas, in terms of accessibility, proximity, use and treatment of the ground while organizing the categorization of territories within large open boundaries.

The strategic green is structural for Barcelona because it can be used to rebalance density and the environmental aspects of the city. It is an assembly of outdoor areas that make up a system of facilities for rest and leisure. These areas, furthermore, contribute to the improvement and transformation of their more run-down surroundings. The green strategy is planned as a long-term process with heavy public funding through multiple concepts that stand up to both the public and private uses that these large free areas often undergo. After many years of disregard, without any grand attempts at transformation, it is necessary to redefine the planning parameters to control the destiny of these urban environments.

The strategic green spaces stand out in terms of their size and setting, with the city as their stunning backdrop. Some of these spaces have a complicated topography that up until now has protected them, being adjacent to marginal residential areas and having irregular qualities. Until 1992, their connectivity to the entire city was deficient, which explains the lack of planning, investment and attention that for many years they had received.

The green strategy approach is based on a careful recognition of their physical, social, historical and structural characteristics, and the pattern for revitalizing them is similarly specific in proposals for action. To define their use and treatment, falling between urban parks and natural ones, the role that each will play in the city, comprehensively and locally, must be determined. These very localized approaches need to be set limits that are not simply linear, but are defined by multiple criteria taking into account the boundaries and elements of transition between the dense urban fabric and the empty areas. Such studies allow for definitions of access, proximity and complementary use of the strategic greens by the surrounding neighbourhoods.

FOUR EXAMPLES OF A FAR-REACHING STRATEGY

Strategic-green actions concentrate on four very different areas of Barcelona. Each one has its own specific features and its own impact on the transformation of the city. The supporting studies are at different stages of completion. Aside from the global studies, the plans and projects for each area are in line with the strategic aims.

ORIOL CLOS

Collserola is part of the metropolitan park that makes a bridge between the dense urban fabric and the nature area managed by the regional government. The relationship is solved with a complex line consisting in a transition strip of varying widths and a variety of features. Given the heterogeneous nature of the locations and the morphological relationships, the strip is structured with open areas that are treated as parks without breaking continuity with the large protected forest it embraces. The elements of continuity, the crest and the connecting green corridors help to set the scale of the whole, as opposed to the urban sprawl. The mountain can be seen from many places in the city. The criss-crossed visual relationships between the mountain and the metropolitan park enhance the park's value as a point of reference along the coastline. It mirrors the way Barcelona opens out to the sea.

Tres Turons is a large empty area covering three hills inserted in the densest part of the urban fabric. It can be conceptualized as an observation balcony that looks out in all directions from the highest part of the city. It is little known to most of the inhabitants of the city. The area is interspersed with buildings that were pushed into the edge of the empty area with the transformation of the city. They form pockets of residential fabric that are continuous with existing neighbourhoods. The strategic-green plan here will enable certain deficiencies of the urban context to be addressed, such as providing community facilities and more points of access to the park. The area as a whole will consist of planting forests that are easy to care for, organized paths to connect the adjacent neighbourhoods and viewing points at the highest points. Gaudí's Park Güell lies within Tres Turons. There is a master plan to enlarge it and organize the surrounding service facilities that would protect the monument site, which is currently under enormous pressure.

Montjuïc is a large central park in Barcelona undergoing transformation. It is situated on a mountain overlooking the port. Originally a military site, it has gradually become integrated in the urban framework with a series of urban planning works, such as the universal Expo in 1929 and the Olympic Games in 1992. The strategy poses two lines of action: to reinforce the park's central location by increasing points of access from every side and to enhance the value of its natural, botanical, geological, sporting and cultural wealth. This involves plans to open up new connections, define and distribute compatible use, address public and individual mobility and promote diversity. The plans for action are stratified in three levels. The first level involves improving contact with the city, which is very densely equipped with facilities and has a formal layout. At the second level is an intermediate strip that will be prepared for the new elements that will lend cohesion to the whole and at

Figure 2.5 Collserola.

Figure 2.6 Tres Turons.

the top the higher ground will be treated as an open nature area around the castle monument.

Ciutadella is Barcelona's historical park. It was built in 1888 on the site of the old military citadel. It is an enclosed park. Up to 1992 it constituted the eastern edge of the city centre, together with the railway. Inside the park are very large political and cultural facilities, as well as the zoo. The latter is being remodelled and part of it is being moved to another area. This huge open area of formal and equipped parkland is being opened and extended and recognized as a link between the historical centre and the renovated fabric in the eastern part of the city, the Olympic village and 22@ in Poblenou. The elements stressing continuity include a structural unit, facilities and buildings that are representative of the city as a whole. Thus, the park is becoming a symbol of the city's renewal by accumulating a succession of significant urban elements in a single spot. Building on a historical foundation with a forward-looking approach, and using the compactness that is recognizable in contrast with empty areas, define the strategies for the transformation of Barcelona.

Figure 2.7 Montjuïc
(Martin-Mas-Mas, architects,
2007).

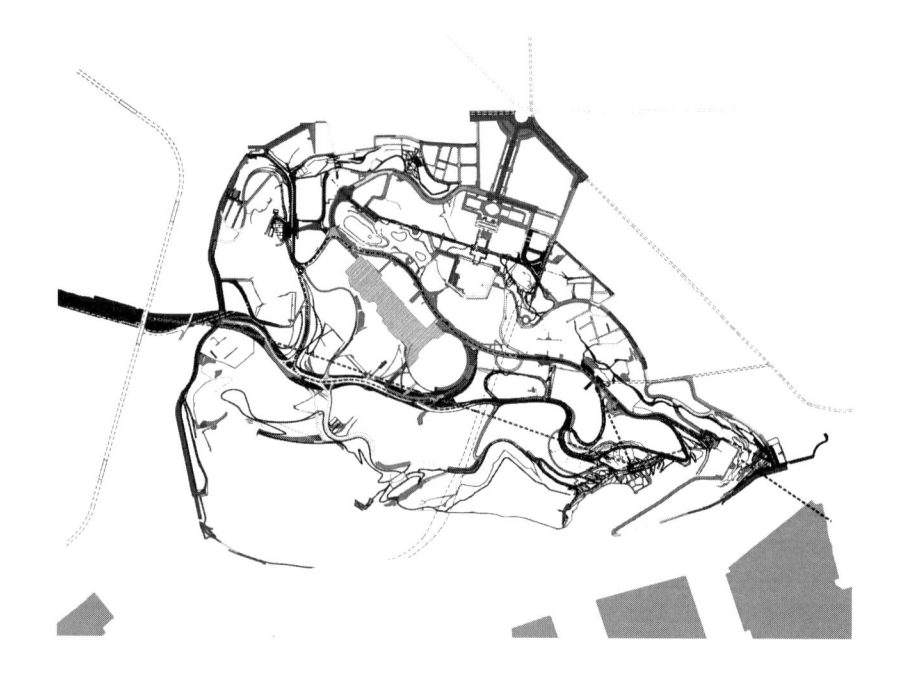

Figure 2.8 Ciutadella
(Batlle-Roig, 2003) (Note: for
Figures 2.9–2.15 please see
colour insert 1).

3 FROM INDUSTRIAL CITIES TO ECO-URBANITY
The Melbourne case study

ROB ADAMS

INTRODUCTION

Many cities have actively pursued and implemented urban design strategies to arrest their decline. These include the acupuncture approach of Bilbao, the public infrastructure of Bogotá, the public spaces and social change projects of Copenhagen, the added cultural institutions of Temple Bar, Dublin, height limits in central Berlin, the sustainable agenda of Bo01 in Malmö and the city-wide strategy of Melbourne. The results have seen an improvement in the social, economic and environmental indicators for these cities. Bilbao has seen an increase in visitors that has made it necessary to expand its airport to handle up to three million passengers a year. The upgrade of its metro reduced travel time by 22 million hours and reduced the number of cars entering the

Figure 3.1 Bilbao, Spain.

Figure 3.2 Community hall in Copenhagen, Denmark.

city a day by 9,000 a day. Bogotá increased its green space per person from 3.2 m^2 to 5.9 m^2, bicycle usage increased from 0.5 per cent to 5.0 per cent, while 80 per cent of the population now use public transport. Over the past decade crime has dropped by about 70 per cent and kidnappings by 87 per cent.

Melbourne has turned its central city from a mono-functional business centre into a multi-functional activity centre that has on numerous occasions won the most liveable city rating worldwide.

FROM LIVEABILITY TO SUSTAINABILITY

The turning point for Melbourne came in the 1980s when its citizens spoke out about the slow destruction of their city. Inappropriate international style developments, the invasion of the automobile, destruction of heritage areas and the decline of the central city saw new political forces emerge at both state and local-government level: their success at the polls allowed them to reset the agenda for Melbourne. A simple vision to transform Melbourne's ailing business district into a central activities district, while retaining the physical characteristics that made it distinctively Melbourne, was adopted and incrementally implemented over 20 years.

Figure 3.3 Temple Bar, the cultural quarter of Dublin, Ireland.

Since the 1980s both state and local government have pursued a strong city-improvement agenda, and while both levels have not always appeared to move in unison, the reality is that there has been a high level of cooperation. These include a joint vision to improve Melbourne's relationship with the Yarra River, shared planning powers and partnerships on key projects such as returning residential use to the city using the Postcode 3000 project, the closure of Swanston Street to through-traffic, and the development of the City Square, Federation Square, Burrarung Marr, a new 8 hectare park, Docklands and a new Plenary Hall. All these have combined to bring significant improvement into the central city.

DESIGN PHILOSOPHY

While both levels of government developed parallel strategic documents, the articulation and implementation of the detailed design agenda fell to the City of Melbourne. This design philosophy, outlined in the 1985 Strategy Plan, was both modest and simple, possibly reflecting the limited resources available to the City at this time. In essence it called for the City to build on its existing strengths in a manner that reflected the local character. It also called for a turn-around strategy and for the proactive increase of uses in the central city, namely turning it from a central business district into a

Figure 3.4a Historic images of the north bank of the Yarra River showing the naturally occurring Turning Basin (a), its removal through land reclamation (b), and the concept plan to restore the Turning Basin (c).

Figure 3.4b.

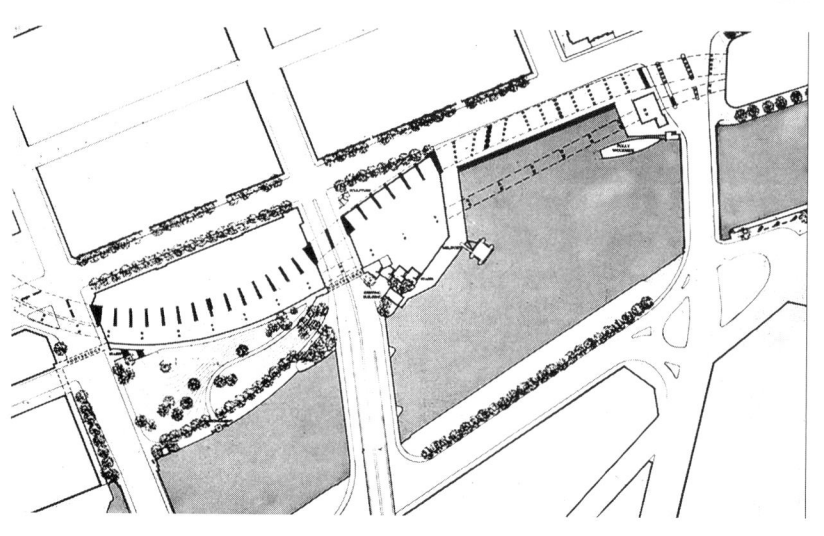

Figure 3.4c.

Figure 3.5a Batman Park, c.1980s–1990s and the restored Turning Basin, c.2000. These 'before' and 'after' images show restoration of the Turning Basin (a) and replacement of roadway with public open space (b).

Figure 3.5b.

central activities district. This would be best achieved by reintroducing a residential community.

The city's existing strengths and physical patterns were identified in the strategy plan and later elaborated upon in the publication *Grids and Greenery*, published in 1987. This document provided a vision for the future of Melbourne. It told the story of Melbourne's urbanization, laid down generic urban design principles and defined elements and relationships that characterize central Melbourne. It described city form in terms that Melburnians recognized and understood, showing how simple things like streets and boulevards, waterways, parks, transport infrastructure, the city centre and heritage built forms interact to create familiar yet distinctive city features. Today, this analysis of Melbourne's urban form is just as well-placed as it was in 1985 to act as the driver for the city's continuing growth and revitalization over the next two decades. In an era of rapid change, with subtle modifications appropriate to a new social, economic and environmental context, the 'enduring assets' remain remarkably stable.

There are many examples of the incremental approach to the improvement of the city's major physical patterns, many of which are recorded in the *Places for People* document produced by the City of Melbourne and Gehl Architects in 2004. The story of the transformation of Melbourne over the last 20 years, it can be structured into five parts: local character, density, mixed use and connectivity and the high quality of the public realm. In illustrating how Melbourne has developed these aspects of its city, this chapter concludes by covering some of the economic, social and environmental outcomes that have evolved from this approach.

LOCAL CHARACTER

As cities move towards globalization there is increasing pressure from communities to protect their local identity so as to retain a point of difference. It is also becoming more apparent that in retaining and supporting local character, cities often retain and support local skills and materials. In Melbourne during the late 1970s, there was increasing concern about the gradual loss or modification of the city's heritage assets in both buildings and streetscapes. In responding to this, the City undertook a comprehensive survey of all its heritage assets and developed a system of ranking both buildings and streetscapes. For the first time the City was able to give a consistent response to developers or future owners as to the likely constraints covering the redevelopment of the properties.

In the 1985 Strategy Plan, this led to parts of the city being categorized as either areas of stability, that is, where heritage was a major consideration, or key development areas where development would be encouraged. This simple mechanism gave guidance to the market and allowed the City to become an early player in the development cycle. Heritage controls have been essential aids to retaining the historic amenity of much of the city and have, over time, contributed to a strengthening of local character and economic stability.

Alongside heritage buildings and streetscapes, the grain and small size of a city's morphology is an integral part of its character and feel. In Melbourne's case, the small-scale subdivisions contained within the height limits of the central area and penetrated by north–south lanes and arcades is a characteristic pattern of central Melbourne. Defending this against the twentieth-century development patterns of consolidation and sameness became a major challenge for Melbourne. The Australian passion for landownership and rights of development made the introduction of planning controls to prevent the consolidation of sites particularly difficult. The solution for Melbourne needed to be channelled through market forces if it was to be effective. The reintroduction of residential use into the central city provided the opportunity to recycle small to medium buildings for inner-city residential use. These buildings were redeveloped with multiple ownerships securing their long-term stability. Scattered throughout the central city, they effectively put a halt to large-scale consolidation in the historical core of the city. This rapid process in the last decade of the twentieth century not only restored a residential population to the central city but protected the lanes, arcades and small buildings that are the source of Melbourne's character.

While conservation controls are a key tool in the development of local character, one of the most powerful urban design strategies is the effective use of development controls. In the case of Melbourne's centre, the retention of height controls, accompanied by a requirement to build up to street frontages and to provide a 75 per cent active frontage onto the main streets has, over 20 years, slowly reinforced the city's strong streetscapes. One of the most successful strategies is when clear, concise development control frameworks have been put in place before the development process commences. Such an example can be seen with the Commonwealth block plan development for the Telstra site in Exhibition Street. In this case, a concise document produced before the architect's brief was finalized allowed the City and the future developer to work closely together, thus minimizing disruptions through the approval process. The result was the retention of a quality streetscape incorporating a number of heritage buildings. It also provided the architect with relative freedom in the design of the office tower.

There is a strong argument that simple mandatory development controls are more effective than performance-based controls and in most cases provide better urban design outcomes. This is to be seen in many European cities such as Barcelona, Paris, Prague and Berlin, where the effective use of fixed height limits has resulted in high-quality streets and public spaces, and, as will be seen later, has accommodated both high densities and mixed-use outcomes.

The beneficial effects of implementing a height limit seem to work in the following ways. They clearly help to establish land values and therefore guard against inflated speculative land prices that force developers to seek greater development potential than a site can comfortably accommodate. They encourage a built form out to the site boundaries with internal courtyard spaces, lanes or arcades. Not only is this built form very efficient in generating usable floor area, it also usually results in a clear distinction between the public and private realms in residential accommodation, with living rooms facing streets and bedrooms facing the inner courtyards. This allows inner-city residences to work well with the inevitable noise and activity of a successful city. They provide equitable access to solar energy, which will become increasingly important as energy prices rise.

While conservation and development control were immediate tools that could be applied to assessing one-off development proposals, large areas of central Melbourne had become degraded or were in need of gradual change or repair. To this end the City developed and put in place a series of master plans that could be incrementally implemented to ensure the return of special areas of the city. In the main, these consist of parks and gardens, but include areas such as the river, which had taken on the role of an industrial drain so that the city turned its face away from its banks.

The master plans tended to concentrate on the physical aspects of such areas and allow for the forward budgeting and allocation of resources that otherwise might be formulated on an ad hoc basis. Much of the expenditure on many of the city's main projects was not funded from the Council's budget. The existence of a clear philosophy outlined in *Grids and Greenery* and supported by achievable master plans, has enabled the easy and quick expenditure of monies sourced from other private- and public-sector partners. The master plans have also facilitated the consistent delivery of major projects that extended over many years or decades, such as the north-bank redevelopment in central Melbourne.

While many of the actions mentioned here were district-wide programmes, the master plans tend to be very area-specific and are designed to mend the urban fabric. If master plans were successful in dealing with and guiding large-scale interventions, technical notes were needed to rein in the daily invention of new and ever more exotic details used at street level. In 1985 technical notes were introduced by the City to ensure consistent treatment in many of the day-to-day details used throughout the city area. These covered such things as street trees, street furniture, kerbs and paving details. The early development of these details has allowed a consistent approach on all projects and the easy dissemination of information to developers and the community. It has also allowed the City to produce a consistent range of street furniture that now generates an annual income from other municipalities registering to use its designs. The consistency of treatment experienced internally has also eliminated much abortive work between the construction of a concept and the creation of the engineering plans for construction. Additional notes are produced as required and included in the series, and the publication has moved from a paper-based system to an online system.

DENSITY

There are many reasons why cities need to increase their density. They need to reduce the consumption of land required for agriculture, reduce travel distances, improve the use of expensive infrastructure and maximize knowledge. They also need to improve the public's safety through passive surveillance, enhance vitality and promote viable public transport, as well as reduce energy consumption. Like many post-industrial cities, Melbourne suffers from very low densities and until the 1990s the city centre lacked a significant residential population. Since the 1980s a number of strategies have been used to help increase the densities downtown. By far the most successful over the last 20 years was the introduction of Postcode 3000 in the early 1990s. This

FROM INDUSTRIAL CITIES TO ECO-URBANITY

1836
city-block plan

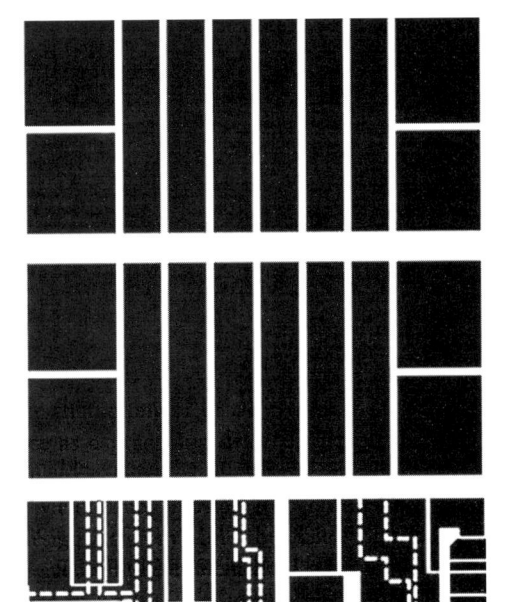

Nineteenth-century
subdivision

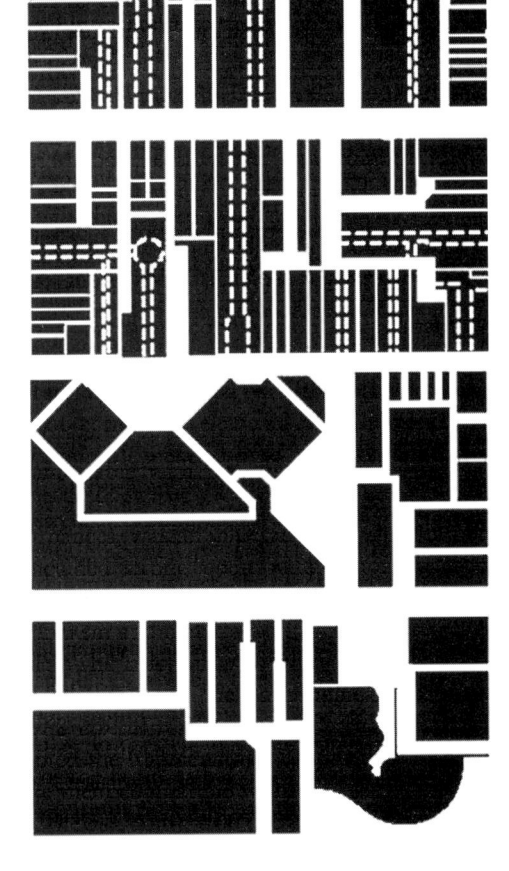

Twentieth-century
consolidation of land

Figure 3.6 The consolidation of blocks during the 1960s–1980s, such as at Collins Place, resulted in the loss of laneways and fine-grain street façades. This is now being reversed through the City of Melbourne's Active Frontages policy (Note: for Figures 3.7–3.14 please see colour insert 1).

programme, designed to reintroduce a residential population into the central city, was spectacularly successful. Using a suite of incentives such as changed regulations, financial assistance, improved street-level environments and promotion, the City managed to reintroduce over 30,000 residential units in just over 15 years. This programme not only led to the reuse of existing under-utilized building stock but also to the redevelopment of under-utilized land close to the central core, such as Docklands.

A challenge for most ex-colonial cities is the tyranny of low density, with housing and facilities spread thinly over vast areas of the metropolis. The solution is to take the 5 per cent of land surrounding rail stations and bus stations and increase their densities to over 150 people per ha, and to do this in a manner that produces quality mixed-use sections of the city. It is not a case of plumping for suburbia or for urban density but rather of providing controlled urban density around rapid-transit routes, supported by suburbs. The challenge is to take the best of both patterns and lock them into a supportive structure but break the need for excessive travel by private car. This strategy has been adopted by the Victorian State Government in its *Metropolitan Strategy 2030* in which distributed activity centres based on rapid-transport routes are being encouraged. The aim is to follow a similar approach to that undertaken in central Melbourne and thus form local, more easily accessible centres within the surrounding suburbs.

MIXED USE

In addition to density, mixed use is one of the cornerstones to healthy, vibrant and sustainable communities. There are many benefits to mixed-use areas. They include their potential to provide a local skills base, to optimize the use of public infrastructure such as parking facilities in accordance with differences in temporal demand, to increase the viability of local businesses, reduce dependency on cars and to promote walking, convenience and increased personal safety.

Central Melbourne was by the 1980s becoming increasingly monofunctional. Retail, lost to suburban shopping malls, was on the decline; residential use was almost non-existent with only 700 central-city dwellings by 1992; entertainment and leisure activities were on the decline with many of the old theatres closed or closing. In 1985 the Council symbolically changed the name of the central business district to the central activities district and began encouraging greater diversification in the centre. Alongside the successful Postcode 3000 programme outlined above, the city developed retail and events strategies. The combination of all these strategies produced a catalytic response to inner-city vitality that saw 1,500 new bars, cafes and restaurants, numerous supermarkets and 400 sidewalk cafes open up downtown. This achieved one of the City's key objectives: changing downtown from a central business district into a central activities district.

One of the most underrated but successful programmes was the city's arts and cultural programme. This diverse programme aimed at sponsoring arts and culture over the widest possible spectrum. The City of Melbourne has the largest arts programme of any local government in Australia. It recognized very early on the importance of 'the creative culture' in stimulating the revival of cities. Other programmes included successful traditional festivals and sporting activities.

By 2006, when it hosted the Commonwealth Games, Melbourne had started to eclipse Sydney as the Australian centre for the arts, culture and events. Melbourne's population was growing at 1,000 people a week to make it the fastest-growing capital city in Australia.

CONNECTIVITY

Where connectivity fails, cities start to fall apart, barriers are formed and neighbourhoods become dysfunctional, disconnected and often hostile to their users. By contrast, good connectivity promotes improved access to local facilities and free movement within and out of the city. It improves land values, reduces vehicle emissions, encourages walking, improves natural surveillance and provides greater choice of movement.

In 1985 Melbourne had many of the characteristics of a car-dominated city. If it were not for the fact that it had retained its tram and rail infrastructure, the city would have been almost entirely dependent on the motor car. Daily, over 28,000 vehicles used its main street, Swanston Street, bisecting the central city and producing a heavily polluted and hostile environment. Both local and state government recognized the need for change. The first major improvement was to build an underground loop to provide three new underground stations. These combined with the existing two major stations to surround the central core of the city with rail infrastructure. To complement this programme the city in the 1980s embarked on a process of gradual pedestrianization of the central city to redress the imbalance between the motor car, public transport and the pedestrian. The expansion of footpaths and the restriction on cars in streets such as Swanston Street has slowly seen an increase of more than double the number of pedestrians, along with the removal of hectares of asphalt from the central city. This, combined with quality pavement finishes, tree planting and distinctive street furniture, has led to a new level of sophistication in Melbourne's streets. More recently, the city has introduced an extensive network of bicycle links that have brought a rapid increase in the number of cyclists. Almost 5,000 people now choose to cycle into the centre of the city each day.

HIGH-QUALITY PUBLIC REALM

One of the casualties of our modern cities has been the quality of the public realm. The push to suburbia and the concentration of the modern movement in architecture on the design of the object rather than the space between buildings has resulted in a loss of understanding and skills in the design of public space, in particular the most important public space in our cities, our street space. The space between buildings has become left-over space, poorly designed, seldom activated by adjacent uses and often dominated by the car. Belatedly, advocates for the public realm in the form of Jane Jacobs, Jan Gehl and others have recognized the importance of the public realm. A

Figure 3.15 Improvements of public realm on Swanston Street.

high-quality public realm attracts people and activities, increases economic performance, encourages new forms of street activity, increases the pride of the community and improves the potential for social engagement and cultural activities. Pride in the public realm can assist in reducing vandalism, encourage tree planting and reduce the waste washed into the storm-water system.

Melbourne by the 1980s was seeing a gradual reduction in the quality of its public realm. Buildings were being set back from street alignments, blank walls were fronting footpaths, footpaths were being eroded and the general level of care and maintenance was poor. Recognizing the problem, the City established an urban-design unit and gave it authority over the quality of the public realm. This small unit started a programme to improve the physical amenity of the city and produced streetscape plans dealing with street tree plantings and open space. A typical example was the central-city tree-planting strategy, which, when combined with bluestone paving, streetlights and other street furniture, combine to give Melbourne's streets their distinctive character.

Working closely with the City's statutory planners, simple principles requiring all new developments to build up to the property line and provide a 75 per cent active frontage have slowly seen the return of quality streetscapes which contribute to the amenity and vitality of the city's public realm. These incremental changes applied through the development control process are highly effective but often neglected by city authorities.

SUSTAINABILITY

With over 51 per cent of the world's population now living in cities and cities producing 75 per cent of the world's greenhouse gases, it is increasingly clear that cities will need to pay greater attention to becoming more successful in producing liveable, sustainable and economically viable built environments. While the five factors outlined above are considered the major components for achieving liveability, they are also the key drivers for sustainability and economic viability. Melbourne has successfully used these drivers over the last 20 years to turn its central city around; significant effort is still required over the remaining metropolitan area if the region is to remain viable in the changing circumstances brought on by climate change and the retreat of the fossil-fuel economy.

As the central city has increased densities, encouraged a greater range of mixed uses, built on its local character, improved connectivity and access for pedestrians, bicycles and public transport, and developed a high-quality public realm, it has become more financially viable and started to reduce its

Figure 3.16 Melbourne's established tree avenues help to define the public realm and create attractive and high-amenity streets.

environmental footprint. Local rates and taxes have declined by over 50 per cent. Property owners in 1996 who had to pay 13 cents in the dollar on the value of their property are now paying only 6 cents in the dollar. The City has also set an ambitious environmental target of zero emissions by 2020 and has already taken a strong leadership position through buying green energy, replacing street lighting with longer-life and more efficient luminaires and instituting extensive street tree-planting schemes, as well as by installing passive energy collectors. It has been paying greater attention to the design of its own buildings, including the design and construction of Australia's first new six green-star-rated commercial office building, CH2.

Melbourne's success over the last 20 years has been due to its ability to set a clear vision with ambitious but achievable targets and then put them on the ground. Where other cities have produced high-quality documents, Melbourne has managed to achieve a high-quality implementation programme. Using a strong tradition of in-house professional skills in all aspects of its administration, it has mastered the art of successful partnerships and directed the resources of other levels of government and the private sector towards its vision. While working from a modest financial base, it has consistently packaged up large and ambitious projects such as Federation Square, QV, Swanston Street, Postcode 3000 and, more recently, the environmental programme. It has also successfully brought on board key partners in the financing and ongoing maintenance of these projects. It has recognized the need for quality design and delivery and the importance of remaining a leader, rather than just a manager, in the art of city making. Its projects have received over 100 awards from architects, landscape architects and planning institutes, and its views and opinions have been sought both locally and internationally.

Just as Melbourne has managed to climb the ladder of liveability, it now needs to become recognized as a leader in sustainability. As an early member of the UN Impact Cities for Climate Change, a partner in the Clinton Climate Change initiative and a champion of numerous environmental strategies it is already well on the way to achieving leadership status. As such, Melbourne offers an important case study of how a city can take control of its destiny and plot a course for the twenty-first century.

BIBLIOGRAPHY

City of Melbourne (1987) *Grids and Greenery: The Character of Inner Melbourne*, Melbourne: City of Melbourne.

City of Melbourne and Gehl Architects (2005) *Places for People: Melbourne 2004*, Melbourne: City of Melbourne.

Department of Infrastructure (2002) *Metropolitan Strategy 2030*, Government of Victoria. Available online: www.dse.vic.gov.au/melbourne2030 online (accessed 6 May 2008).

4 THE SUSTAINABLE CITY AS A FINE-GRAINED CITY

DAVID SIM

INTRODUCTION

My story starts some decade-and-a-half ago, with a visit to an eco-village in southern Sweden. The virtuous inhabitants of 'Sun Village' lived in well-insulated houses oriented towards the sun, they cultivated food in their allotments, used dry toilets, collected rainwater, wove cloth and so on – their eco-lifestyle putting a 'townie' like myself to shame. However on interviewing a few of the residents, I found that the rural eco-life had

Figure 4.1 Copenhagen, Denmark.

its drawbacks. So many aspects of life were not available in the rural idyll that the residents were forced to drive their cars back and forth across the plains of Skåne in search of employment, education, entertainment, goods and services. 'Of course, I have to have a car – how else would the kids get to ballet lessons?', 'We buy our eco-beans in special shop in Malmö', 'I go to an Alexander practitioner in Helsingborg' and so on. Suddenly I realized that, as a carless city-dweller, I was living a different kind of 'eco lifestyle', because all I needed and wanted was within five to ten minutes walk from my flat.

A second epiphany came a few months later on a cheap charter holiday to southern Spain. My unspecified accommodation was in the infamous resort Torremolinos, an intense few square kilometres of accommodation, leisure and entertainment located at the beach. I discovered, however, that the concentration of activity in Torremolinos which, unlike later coastal development in Spain, is based around a traditional urban centre, not only kept most of the tourists contented within its built-up area (since they could get to where they want on foot – and then stagger home) and few stray out beyond the city limits. Just kilometres away I could find large agricultural fields, unspoiled landscape and well-preserved rural villages. Towns and cities, it would seem, could be very good at efficiently and effectively accommodating all the mess of human activity while protecting the greater landscape from unnecessary intervention.

Figure 4.2a Torremolinos, Spain.

Figure 4.2b.

Figure 4.2c.

LONGING FOR A SUSTAINABLE, PEOPLE-FRIENDLY CITY

The German saying, '*Stadtluft macht frei*', literally meaning 'city air makes you free', dates back to medieval times and captures the spirit of longing from the countryside to the city, from impoverished rural life, from serfdom, toil and soil, for the city and the seemingly endless possibility of urban life, a longing that we can imagine has existed as long as human beings have come together in large settlements.

The city is still the place of possibility, and we can understand the magnetic attraction of city today, particularly in the developing world, where they are seen as economic magnets and beacons of hope and dreams. Only recently have we reached the turning point, where more than 50 per cent of the planet's population lives in towns and cities. The mutual benefit for people living side by side in cities is access to a range of shared resources: markets and meeting places, money and employment, and the opportunities for trading goods, services, skills and crafts, as well as making things, sharing ideas and promoting creativity and, most important of all, exchanging knowledge (news, advertising, propaganda and learning at all levels).

The attraction of cities, however, creates a number of challenges. With the ever-increasing percentage of the world's population living in cities, many environmental problems are today therefore also concentrated in cities – problems created, for instance, by the volume of transport, pollution, use of energy, demands on sanitation and the existence of inner-city slums. Moreover, the increasing number of city dwellers puts pressure on urban land, often resulting in an uncontrolled expansion of cities, which will have negative effects on the greater landscape surrounding our cities.

However, while it is important to protect the landscapes surrounding cities, equal attention must be paid to the protection of our cities. If we are to prevent unnecessary urban expansion, urban space must be recognized as a precious resource. This resource must be carefully considered and managed, and different spatial arrangements can be seen to represent opportunities for a more sustainable way of organizing and nurturing human activities. Sustainability is not limited to technical issues such as using eco-friendly building materials, designing and constructing low-energy housing, using insulation and conserving water. The spatial organization of our cities has an equally great impact on the environment and the extent to which they meet the needs of urban dwellers, and it is especially in this aspect of urban design that architects and planners have an important role to play.

An easy and frequently used argument, especially among property developers, is to promote the dense city as the sustainable way forward, with the underlying assumption that an efficient use of urban space is enough. But spatial efficiency on its own often results in more and more standardized solutions. Regardless of local political affiliation, many cities today follow the same strategies to become important economic magnets. The result is that many new urban areas are becoming more and more uniform and corporate in their look. This tendency is unfortunate. Not only are cultural characteristics often lost, these kinds of developments tend to disregard important human needs.

Figure 2.9 Glòries plan, Barcelona (source: City Council Planning Services, 2007).

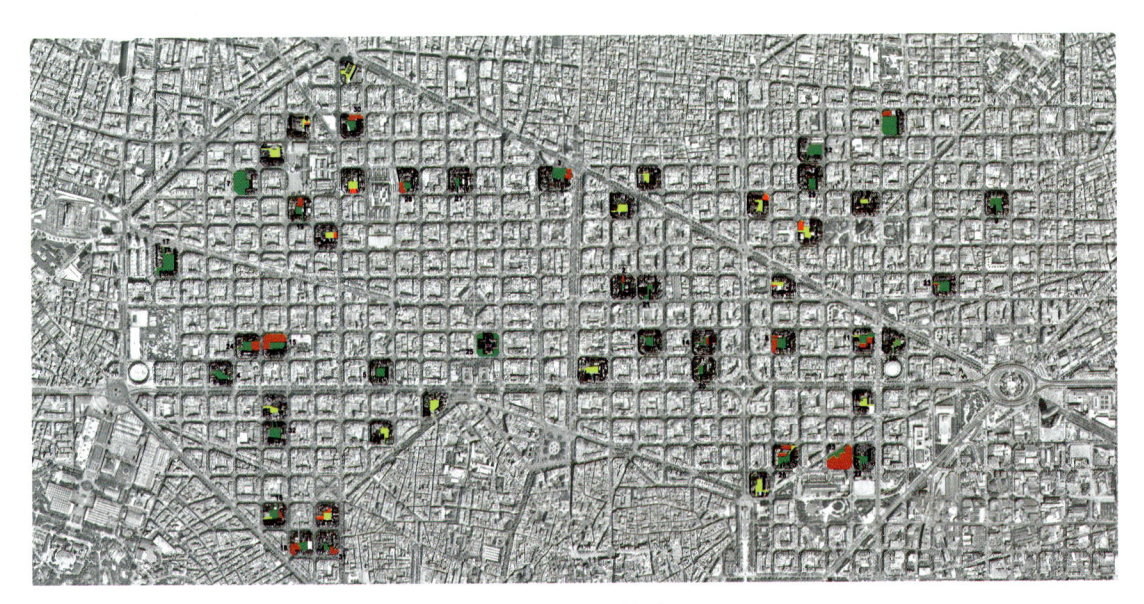

Figure 2.10 New green spaces and local facilities in the Cerdà extension blocks.

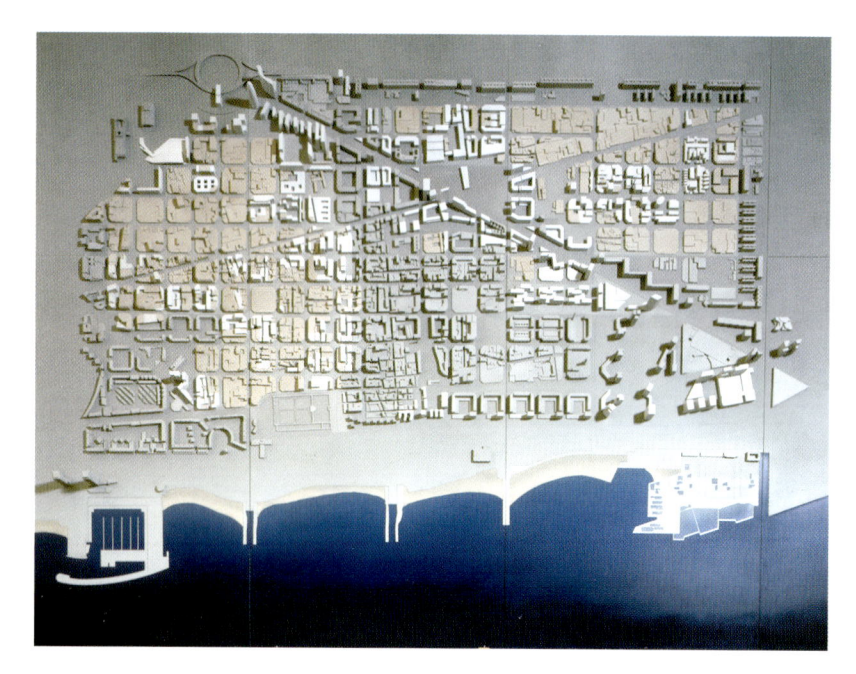

Figure 2.11 Poblenou and 22@ district model.

Figure 2.12 Studies for Montjuic (source: Martin-Mas-Mas, architects, 2007).

Figure 2.13 Mixed-use complex at the Guinardó Market (source: Cantallops, Bayona, Valero, Vicente, architects, 2006).

Figure 2.14 Casemes de sant Andreu plan (source: Manuel de Solà-Morales, architects, 2005).

Figure 2.15 La Marina del Prat Vermell plan (source: City Council Planning Services, 2006).

Figure 3.7 These 'before' and 'after' photographs of Swanston Street and Bourke Street, Melbourne, clearly show the impact streetscape works have had on improving pedestrian amenity.

Figure 3.8a Diagrams illustrating the dramatic increase in residential dwellings in Melbourne's central city.

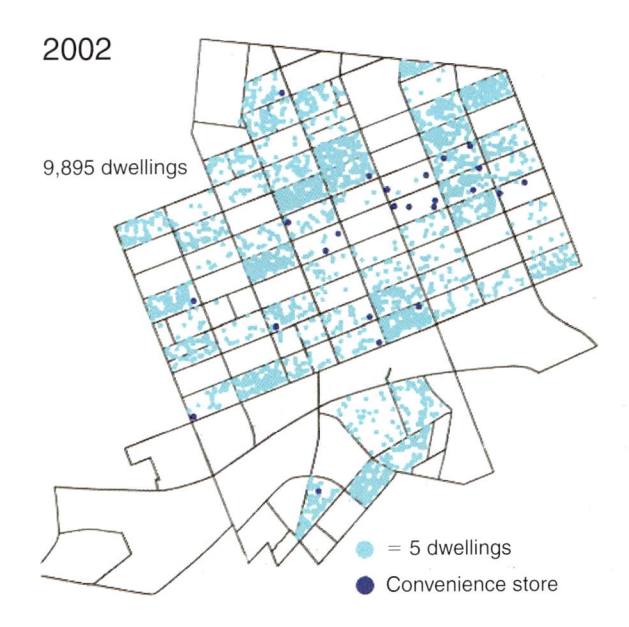

1983

204 dwellings

● = 5 dwellings

2002

9,895 dwellings

● = 5 dwellings
● Convenience store

Figure 3.8b [as above]

Figure 3.9a Outdoor cafés in Melbourne's laneways and the pedestrianization of streets.

Figure 3.9b [as above]

Figure 3.10 Data taken in 1993 and 2004 demonstrate that pedestrian numbers have increased in Swanston Street. This growth is attributed to improvements in streetscape amenity and vitality.

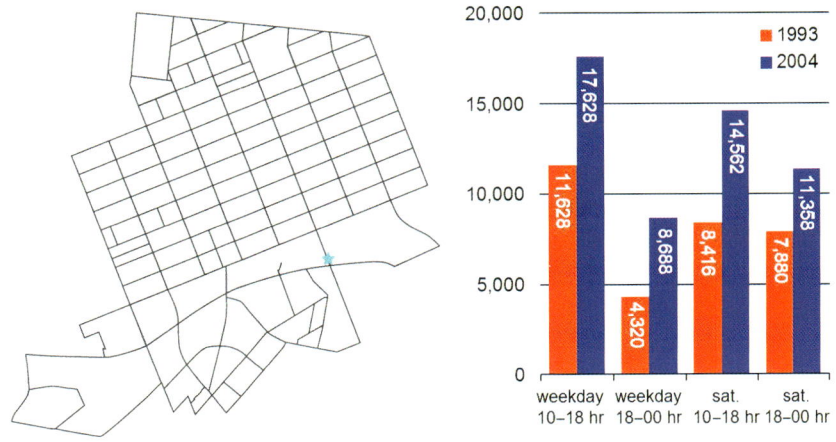

total no. of pedestrians per day – Princes bridge

Figure 3.11 The façade of the T&G building, Collins Street, was incorporated into redevelopment of the site. This allowed the historic streetscape to be preserved.

Figure 3.12a The addition of floors to existing structures allows historic buildings to be retained while ensuring their conversion to apartments is economically viable.

Figure 3.12b [as above]

Figure 3.13a Art installations in Melbourne's streets, laneways and other public spaces.

Figure 3.13b [as above]

Figure 3.14a Art installations in Melbourne's streets, laneways and other public spaces.

Figure 3.14b [as above]

Figure 3.14c [as above]

Figure 3.14d [as above]

Figure 4.4a Grenville Island, Vancouver.

Figure 4.4b [as above]

Figure 4.5 Mårtenstorget market square in Lund, Sweden.

Figure 4.8 Amsterdam 'stoop'.

Figure 4.10c The front porch of the Georgian house.

Figure 4.10d The 'bridge' of the Georgian house.

Figure 4.11b The vestibule of a Scottish house.

Figure 4.12d 'Zaguanes' of the traditional Spanish house.

Figure 4.13b A village in Majorca.

Figure 4.14 The front porch of a house in Tokyo.

Figure 5.1a Hong Kong, one of the world's densest cities, with mixed and intensive land use.

Figure 5.1b [as above]

Figure 5.2 Ibn Battuta Mall, one of a multitude of large, privatized shopping 'experiences'.

Figure 5.3b The Arab Street area of Singapore.

Figure 5.3a The Arab Street area of Singapore, regenerated to a high standard, but almost exclusively the preserve of tourists.

Characteristic of these new urban strategies are building taller and taller buildings with deeper plans and larger floor plates, with fewer entrances and openings between the private and public realm in the name of security and not providing traditional ground-floor amenities such as shops and services. It is not just single-function buildings that are being produced, but whole urban blocks with only one use are being built. These blocks and buildings lack the traditional clarity of front and back, public and private distinctions and whole areas lack legible hierarchies of spatial significance.

Regardless of architectural style, it is the small-scale aspect that is missing in the new urban areas. This means that there is a lack of decoration, finer finish and detailing, especially at ground level, while the smooth treatment of elevations, often by using large curtain walls of reflective glazing, do little to stimulate the senses. Significantly, there is little or no possibility for individual personalization by the inhabitants. These new mono-functional areas are lacking in life at different times of day and because they lack a range of useful services, people are forced to travel to other parts of the city to obtain the complementary facilities to lead a normal life. The lack of enjoyable public space and small-scale detailing makes these places where people simply do not want to spend time. An example of this is the Waverley Steps in Edinburgh. The Balmoral Hotel from 1902 sits on one side of the thoroughfare with the fairly recent Princes Mall (early 1980s) on the other.

Both buildings are complex multi-use corporate entities with a strong need for a clear identity. The newer building presents a long, horizontal, smooth, blank wall to the public thoroughfare with no decoration or embellishment whatsoever. The 45° slope towards the ground plane has the deliberate intention of prohibiting any spontaneous use of the edge, such as sitting down. The older building has highly decorated stonework. In addition there is a vertical rhythm of a series of shop fronts and entrances where the grand old lady hitches up her skirts, revealing these parasite uses that have no direct connection with the activity of the main building. This phenomenon is a kind of urban symbiosis. The 'tenants', the small businesses, get the benefit of a city-centre location and passing footfall and the 'host', the great structure itself, benefits from the busy life around, the safer streets because of constant human presence. Additionally, the host gets extra rental income from the

Figure 4.3 Waverley Steps, Edinburgh.

small businesses, making for a kind of economic sustainability through diversification.

Efficient and effective use of space

Density alone does not make for sustainable and people-friendly cities. We argue for an effective use of the urban space in terms of the city providing density along with diversity, and in so doing increasing the likelihood of proximity to different useful and stimulating activities only by ensuring mixed uses. Ideally, everything the citizen needs would be within 5–10 minutes' walking distance. This proximity saves the city dweller from time- and energy-wasting travel to satisfy everyday needs for life, work, entertainment and recreation. This everyday life that is in walking distance to the dwelling or workplace can in turn increase the likelihood of social interaction, security, cultural richness and positive stimulations for the human senses, all of which make for stronger communities and happier individuals.

THE FINE-GRAINED CITY

The notion of the fine-grained city captures in many ways the vision of a more sustainable and people-friendly city, a vision that takes as its starting point urban space as a precious resource.

Hardware and software

The hardware of the fine-grained city requires a physical structure made up of many independent units, small in size and accommodating a wider range of different activities. This would mean different built structures containing different volumes, floor areas, room sizes and ceiling heights with different natural light and ventilation conditions, different standards of finish and access arrangements. The different units would ideally have different building details (even small variations are significant) and express the work of different architects and artisans. This hardware has to be accompanied by a diversity of functions within an urban area. Diversity is required not only in housing and workplaces, but in fine-grained diversity of different kinds of housing, different kinds of workplaces as well as other functions such as entertainment, culture and recreation. A good example of this type of fine-grained diversity is Granville Island in Vancouver (see Figure 4.4, colour insert 1), where one finds in a very small area such diverse functions as housing, small retail shops, a food market, a cement factory, an arts school, nightclubs, restaurants, galleries, a brewery and theatres.

The diversity must also be thought into the more general nature of spaces. Indeed poly-functional rather than mono-functional spaces allow for a greater flexibility, as does direct access to the main thoroughfare from the building (via shop windows and customer entrances), possible secondary access (for delivery goods, for example) secure storage (such as in private and shared courtyard spaces). This variation would be true not only of indoor rooms but also of shared and private outdoor spaces. Many useful elements and systems for spatial organization can be found in traditional cities and we might use these as clues to new interpretations of urban design. These clues include building elements and systems such as the *porte cochère*, or pend, courtyards and walled gardens, town houses with a cellar, ground floor, *piano nobile* and attic (accommodating many different uses at the same address), outhouses, tenements and single-aspect courtyard buildings.

The software of the fine-grained city would include many different land ownerships in a relatively small area, different forms of tenure and control, including shared equity, cooperatives and very small private holdings, along with pragmatic local decision making allowing spontaneous change of use. Part of the software would also include 'time planning'. A very simple example of enriching the value of urban space is to use the space over a longer part of the day than normal by permitting different uses. An example of this overlapping in time is Mårtenstorget market square in Lund, Sweden (see Figure 4.5, colour insert 1). Here the square is used as a market place in the morning and a car park in the afternoon. In addition, a few times each year it is used as a fairground and for other civic events. The space is poly-functional in time.

Fine-grained networks and public spaces

Also entailed in the notion of the fine-grained city is the city as a web of routes. Rather than a city only of big boulevards suited for fast travel in cars and leading to a separation of the city into different parts that are not necessarily well connected, the fine-grained city is characterized by a network of routes and connections, local paths and shortcuts – useable by cars but much more suited to cyclists and pedestrians. Vital in making the sustainable city is the aim of making daily life in the city more bearable and in doing so reducing the daily need to travel somewhere else.

In many cities we have witnessed the reconquering of public space, a concept that was born in Barcelona in the early 1980s (Gehl and Gemzøe 2003: 18). In the process of developing city centres with a greater focus on public spaces and city life, Jane Jacobs has served as great inspirational source with her book *Death and Life of Great American Cities* (Jacobs 1961). A more varied and complex perception of the city has emerged since then, emphasizing urban space not only for commercial use but also for social and recreational purposes. As a result, pedestrian streets or priority pedestrian streets have become increasingly visible in cities, most notably in European cities. These initiatives help to establish the experience of a fine-grained city, where its many small parts are well-connected and thus more easily experienced. The fine-grained city connotes, in other words, a well-established hierarchy of routes where streets, avenues, boulevards and 'ramblas' along with mews, back lanes and side streets complement each other, as well as different kinds of public spaces, squares, plazas, piazzettas, parks and so on.

Actual smallness

Another important aspect of the fine-grained city is the experience of actual physical smallness – spaces and details which relate to the scale of our bodies and the way our senses work. The fine grain should offer a rich experience with potential variety and a choice of different sensory stimulation when people spend time in the city and allow for a diversity of impressions, activities, functions and people. Smallness represents an opportunity for adding a more human dimension to the urban fabric. This has been an area of concern for Jan Gehl (1971) for many years. He has stressed how our senses for the most part are designed to perceive at a very low speed (a normal walking speed of around 5 km per hour). Moreover, our sensory apparatus faces forward, and our best-developed and most useful sense, that of sight, is distinctly horizontal and considerably wider than the vertical. When looking straight ahead it is possible to glimpse what is going on at both sides within a horizontal arc of almost 90° to each side. This means that the person walking along a street mainly concentrates on the ground floor of buildings on each side of the street and the pavement, and the action taking place in the street space itself. We seldom look up higher than 3 m.

When we are walking at 5 km per hour our other senses work in parallel, making for a highly stimulating sensory, even sensual experience, smelling flowers and plants, perfumes and food cooking; hearing and overhearing conversations, music, the noise of activities, water, birds singing, leaves rustling (though these all requiring a low ambient noise level); touching things within arms' reach plants, materials, textures, patterns and decoration, text and graphics.

MAKING A FINER GRAIN

In our search for the fine-grained city we can learn from other places that have fine grain as a starting point or, perhaps more interestingly, that have adapted a more coarse grain to a finer one, acknowledging the aspects that make places both more useful and more enjoyable. Examples that immediately come to mind of interventions for making a finer grain are limpet or parasite structures that have been added to break down an element that has proved to be too large for the urban surroundings. Such interventions can be seen in the long stretches of stone wall along the banks of the Seine in Paris that have been colonized with the well-known *bouquinistes*. These architectural limpets provide a useful function and enliven an otherwise tedious experience.

Also in Paris, and with a similar enlivening effect, on the Rue du Faubourg Sainte-Honoré, small additions of tiny houses containing shops fill up the leftover triangles of space between the church's rectangular plan and the loose grid of the medieval city.

Having outlined some of the overall principles for a fine-grained city, the remaining part of the chapter will focus on a more specific aspect: what we choose to call the thick thresholds between the public and the private spaces in the city.

THE SUSTAINABLE CITY AS A FINE-GRAINED CITY

Figure 4.6 Stretches of stone wall along the bank of the Seine, Paris.

Figure 4.7 Boulevard St Sulpice, Paris.

ZOOMING IN: THICK THRESHOLDS IN THE FINE-GRAINED CITY

City dwellers adapt to their surroundings, making use of the different spaces – indoors and out, public and private – to make them more suitable for their immediate needs and requirements, as well as more appealing and enjoyable. Among the many spaces inhabited, it is in particular the threshold between public and private that seems to be consistently given attention. Perhaps this distinction is so consciously attended to and managed because it seems to users like the ultimate threshold – that between the household and the rest of the world. Thus, many of the examples that are described below are thresholds that are closely related to and add a transitional zone between the home and the public street.

The human need to establish different and often subtle layers between the private and the public seems to constitute a universal phenomenon. These layers must take very different forms and cultural expressions, but one could argue that the creative ways of establishing them represent the universal comfort derived from small details and a transitional zone between the private and the public. We call these layers 'thick thresholds'. The examples are numerous and represent a wide range of arrangements, which may or may not be officially planned. The first obvious example is the Amsterdam 'stoop' (see Figure 4.8, colour insert 1). To begin with, the Amsterdam stoop had the practical function of giving access to the dwelling over the above-ground basement, but the transition zone that the steps created soon became a place for spending time, a place that is physically advantageous in relation to the street and offers good views as well as a feeling of protection. The stoop also became a place to express pride, for example cleanliness, and identity, with decoration and personalization.

Later, the stoop crossed the Atlantic to New Amsterdam (New York), and in a new climate, a new scale and new sociocultural situation became an even more significant place to spend time. The density of New York city, both in terms of inhabitation of buildings and footfall on the streets, as well as the presence of people from different cultures (such as the Mediterranean basin rather than the Anglo-Saxon north) where spending free time outdoors is more common, all make the stoop a useful space to inhabit for the same reasons as those in Amsterdam.

An example Jan Gehl (1977) gives is the veranda of the traditional Australian terraced house. This climatic adaptation of the English terraced house has proved to be hugely significant in establishing an inhabitable space on the threshold between public and private. This is a place specifically intended for spending time, suitably dimensioned for furnishing with a table and chairs for taking meals and refreshments and thus allowing significantly more time to be spent there. More time spent on the veranda means more opportunities for people-watching and interaction with other people.

The veranda, or front porch, is also found in Canada, and Jan Gehl has many anecdotes about living in a street with verandas in Toronto. The less physically active, particularly older people, are able to partake in the life of the street and at times be and feel useful, helping them to be socially included.

Figure 4.9 Examples of verandas.

When they are at home during the day when other people are away at work or school, they watch comings and goings along the street, building up the knowledge of the local residents and their lives that makes for a stronger community. Apart from their spontaneous conversations with passers-by, the street watchers are important for security, by recognizing strangers or unusual activity, thus preventing crime and vandalism, and quickly helping well-intentioned strangers like people visiting friends and making deliveries (see Figure 4.10c, colour insert 1).

The complex and strict social stratification in the British Empire produced a very particular urban residence, the Georgian house with a sunken forecourt and little 'bridge' to the front door, as found in Edinburgh's New Town (see Figure 4.10d, colour insert 1). Although this was a complex solution to servicing a large household consisting of the owners and their servants, in later times both the thick threshold of the forecourt and the bridge have been claimed as gardens, markets stalls, storage spaces and places that can be personalized. The multiple entrances, originally intended to distinguish owners and their visitors and servants, service-workers and delivery-men, have permitted the subdivision of these buildings into dwellings more appropriately sized to today's needs and budgets. This subdivision allows different socio-economic groups to live at the same address, albeit in different spatial conditions.

Another Scottish example is the vestibule or lobby, which is more than a mere buffer against the climate. The play between the heavy outer front door and the inner glass door (often decorated) can express a range of different invitations to the outside world by degree of openness. Personal possessions exposed in the lobby demonstrate the owners' trust that they will not be stolen, and by not stealing them passers-by affirm their own respect of that trust. The addition of a small front garden with a gate only increases the complexity and subtlety of this expression as there is even more space for

Figure 4.10a, b Threshold zones, Georgian houses, Edinburgh.

Figure 4.11a The vestibule of a Scottish house (see Figure 4.11b, colour insert 1).

Figure 4.12a, b, c 'Zaguanes' of the traditional Spanish house (see Figure 4.12d, colour insert 1).

Figure 4.13a A village in Majorca (see Figure 4.13b, colour insert 1).

personal expression outside the dwelling and there is another layer of control with the gate.

In traditional Spain, with a strong urban tradition of building, even in small villages, great attention is paid to the entrance of the dwelling. Double sets of doors, porches (or *zaguanes*) and iron gates make the transition between public and private a place of celebration and, as with the Scottish front porch, allow a flexible and ever-changing range of physical and hence social situations to occur. In the Spanish climate, the iron gate maintains physical security while affording the passer-by not just visual interest but a fuller sensory stimulation including smelling food and overhearing conversation and music.

In villages in Majorca that often do not have pavements, householders place potted plants around their entrances, demarcating their territory with a buffer zone that gives them a very small space just outside the very private world of their stone houses. Additionally, the plants soften the hard space of the street.

Finally, to include a non-Western example, it is worth mentioning Tokyo, where the thick threshold of the front porch works in ways that are like the Scottish weather porch and the Spanish *zaguan* (see Figure 4.14, colour insert 1). Japanese advanced technology allows sliding doors, so that the space can change completely in character, depending on whether the doors are open or shut, because they disappear completely and make the spontaneous spatial change total. Additionally, we also found personal storage solutions (seen in the British house 'bridges') as well as the plant-pot zones, developed even more than in Majorca, which spread beyond the property line and illegally inhabit the public territory.

The contributions of thick thresholds

On a small scale many of the thick thresholds fulfil the need for people to spend time on the edge between public and private worlds. Inhabiting this interface – hanging around, standing, sitting, watching – provides an opportunity to see and interact with neighbours and passers-by. The security of physically being on one's own territory makes it easier to dare to engage with the surroundings. Additionally, having a legitimate excuse to be on the threshold – cleaning, tending plants, checking to see if there is post – give greater confidence to the urban dwellers to spend time there.

Another phenomenon of the thick threshold is people placing their possessions on and along the interface between the public and private realms. There are obvious practical reasons for keeping possessions intended for use outside (such as shoes, umbrellas, bikes and sports equipment) near the door, and washing dries faster outdoors than in. Additionally, using the threshold in this way allows for the possibility of using them more spontaneously and therefore more frequently if, for example, your skateboard is kept just by the door. There is a natural desire to linger at the edge and watch, and a need to mark and celebrate territory. Ultimately, all of these acts civilize the public realm, making life in the city better, more bearable and ultimately more enjoyable.

The personalization of space outside a dwelling with possessions establishes a sense of belonging and intimacy, which is both immediately physically sensually stimulating as well as mentally stimulating (with triggers and associations such as seeing toys, sports equipment and clothing). The physical presence of private possessions and the signs of care and embellishment work together to make a place feel safe and secure, even when there are no people there.

While some scholars have argued for a clear demarcation between public and private properties as prerequisite for a lively and safe street (Jacobs 1961, Newman 1972), the thick thresholds described above help to establish the city as a safe space even more effectively than strictly defence-oriented spatial arrangements. This is because the transition zones allow for a more enriched townscape and spatial identity, where people are attracted by the idea of spending time in the streets, resulting in public spaces becoming social spaces with many more eyes on the streets. This enhanced sense of security may also bring a potential for conflict between people with different sets of values who have been brought together in the same place, as it may be that greater physical diversity promotes greater conflict-ridden sociocultural diversity. But at the same time the threshold spaces and the tone they set create spaces for negation and conflict resolution, for compromise and tolerance.

A step backwards or the way forward?

As the different examples illustrate, the thick thresholds can be found in many forms and at many levels. As a way towards a more fine-grained sustainable city, the challenge is how to introduce the thick threshold as an integrated part of planning. Some might argue that the notion of the fine-grained city echoes the traditional city that emerged in Europe in the Middle Ages, characterized by streets adapted to foot traffic and squares tailored for necessary uses such as markets, town meetings and religious processions. Here, the micro-climate was adapted to make it as pleasant as possible to spend time in these spaces, as almost every aspect of life (except eating and sleeping) took place outdoors. In the traditional city we find networks of often narrow and winding streets with many small details along the way, as a result of the different uses that the streets and squares must accommodate. With a strong emphasis on supporting the comings and goings of pedestrians, the traditional city is simply characterized by a more human scale than many new urban areas. Among these human dimensions are more varied layers between the private and the public.

The scale of the traditional city is very different from that of new urban areas. Traffic planning and accommodating cars have been given greater priority than accommodating people, and many new urban areas are mono-functional and quite uniform and corporate in character. Today, contemporary apartment blocks and office buildings generally contribute very little to their surroundings. They are highly introverted, and their design sharply demarcates areas of privacy and enclosure from the surrounding environment. Instead of active ground floors, many buildings in the new urban areas suffer from frontages that are either completely closed off with a solid wall or sealed in with reflective glass. New cities do not need to look like old ones. However, more than ever, new cities need to work well for pedestrians and need to reduce the need for motor transport. Cities today need to be flexible and adaptable to accommodate demographic and economic change. Our vision is a fine-grained city that takes its inspiration from the traditional city, but only to serve as principles for the planning and design of new urban areas. We are not proposing a new urbanism, because this is not a question of architectural style. We are proposing something more like 'real urbanism'. Indeed, there need not be a contradiction between the fine-grained city and new urban areas.

CONCLUSION

The need to protect agricultural and natural landscape from unnecessary expansion seems obvious. The idea that cities are and can be efficient and effective systems for accommodating the wide range of human needs and desires seems sensible. However, to make the city really work well, it must appeal to people at all levels and accommodate their constantly evolving needs and desires. We see the fine-grained city as the key to new development as it can adapt and change to accommodate all that human life entails. Many of the clues to urban design solutions can be found in the physical structure of existing fine-grained cities and in the way their citizens inhabit their cities. The city has to be truly functionally efficient – everything you need should ideally be accessible within a few minutes' walk – in order to be socially effective. It should offer attractive choices and alternatives in everyday use so that the inhabitants never need to get bored. The city should stimulate our senses and our minds and offer us all the opportunities to engage with life in all its aspects. Only then can the city be a sustainable container for life.

ACKNOWLEDGEMENTS

This chapter is partly based on research carried out by David Sim and partly based on theories developed by Jan Gehl and, later, Gehl Architects. The former involves studies on proximity and the thick thresholds carried out at Lund University. The latter includes theories of how the human senses inform our experience and use of the urban space developed by Jan Gehl during his time as professor at the Danish Architectural School and later transformed into more concrete planning principles for enhancing urban life at Gehl Architects. Few works have only one author. A strong spirit of collaboration pervades Gehl Architects and the author gratefully acknowledges the collaboration of colleague Louise Kielgast in preparing and writing the original symposium paper as well as in developing and pre-editing this chapter.

BIBLIOGRAPHY

Gehl, J. (1971) *Livet mellem husene*, Copenhagen: Arkitektens Forlag.
Gehl, J. (1977) *The Interface between Public and Private Territories in Residential Areas: A study*, Parkville, Victoria: Department of Architecture and Building, Melbourne University.
Gehl, J. and Gemzøe, L. (2003) *New City Spaces*, Copenhagen: The Danish Architectural Press.
Jacobs, J. (1961 [1993]) *Death and Life of Great American Cities*, New York: Random House.
Newman, O. (1972) *Defensible Space*, London: Macmillan.

5 FROM THE COMPACT CITY TO THE DEFRAGMENTED CITY
Another route towards a sustainable urban form?[1]

MIKE JENKS

In Western nations there is a comforting notion that, at least, an answer has been found to achieving sustainable urban forms. The paradigm that has been promoted is for the high-density, mixed-use, transport-efficient and socially and economically viable built form. This has been embodied in European policy since the early 1990s (Commission of the European Communities 1990) and numerous examples of 'sustainable' environments have been built in the United Kingdom, Europe and the West (Commission for Architecture and the Built Environment [CABE] 2008). One of the dominant concepts behind these policies and examples is that of the compact city – claiming to satisfy the triple bottom line of sustainable development: environmental, social and economic sustainability.

A number of publications, guidelines and polemics, attempt to define and characterize the form of the sustainable city, and to determine which forms may most affect sustainability. But this is a complex issue. The physical dimensions of urban form may include its size, shape, land use, configuration and its distribution of open space. However, sustainability depends on the more abstract issues mentioned above, and there appear to be not one, but a number of urban forms that may be sustainable. Yet, much of the debate about the sustainability of cities and urban forms has focused on increasing the density of development, ensuring a mix of uses, containing urban 'sprawl' and achieving social and economic diversity and vitality – often characterized as the concept of a 'compact city' (Jenks *et al.* 1996; Williams *et al.* 2000).

In the United Kingdom, government policy embodied such principles through an Urban White Paper in 2000, mostly based on the report of the Urban Task Force led by the architect Richard Rogers and numerous planning-policy guidance documents (Department for Communities and Local Government [DCLG] 2006a; Department of the Environment, Transport and the Regions [DETR] 2000; Urban Task Force 1999). In the United Kingdom this policy is being implemented in many towns and cities. It aims at more compact, high-density and mixed-use urban forms in the belief that they will be sustainable (DCLG, 2006b).

With the implementation of this policy, practice has, to some extent, overtaken the knowledge and evidence needed to assure its success. Notwithstanding examples of good practice, advocacy rather than research has often characterized the debate and has led to many claims for the influence and benefits of different urban forms.

The European Commission in 1990 was an early and influential advocate of urban containment and more compact forms. The hypothesis was that compact urban forms would reduce urban sprawl, protect agricultural and amenity land and lead to the more efficient use of existing, previously developed urban land. With a mixture of uses in much closer proximity, alternative modes of travel would be encouraged such as walking and cycling, and public-transport use would also increase. This in turn would lead to environmental, social and economic benefits. Not only were these ideas embodied in UK government strategy, but similar concepts to achieve urban sustainability can also be found, for example, in the United States with new urbanism and smart growth initiatives (Katz 1994; Smart Growth 2008). All tend to advocate contained high-density and mixed-use urban forms that reduce travel distance and dependence on private transport, as well as being socially diverse and economically viable.

The policies and solutions proposed implicitly, and sometimes explicitly, argue that there will be benefits in terms of environmental, social and economic sustainability. The effect has become evident. In the United Kingdom over the past five years there has been an increased take up of brownfield land, fuelled by government targets for building 60 per cent of new homes on re-used urban land (DETR 2001). The claimed 'upsides' may also have matching 'downsides'. A more intensive use of existing land means there is a potential loss of open space and amenity. Environmentally, having less open space is likely to have adverse effects on biodiversity and the provision of environmental and ecosystem services (such as water and drainage), the consequences of which are largely unexplored. Socially, the impact of 'intensification' or compaction may affect the quality of life of users and these adverse effects may in some respects fall unevenly on the poor.

Higher densities are strongly advocated and on the ground there are examples of schemes being built and proposed at higher densities than recommended in the guidelines (CABE 2005). Theoretical studies have been undertaken on physical capacity, showing how reduced parking standards can increase density and examples of 'car free' housing have been built (Canmore Housing Association 2008; Stubbs and Walters 2001). This more intense development may have benefited public transport and access to facilities, but it is not clear that there has been the modal shift in transport use necessary to yield the claimed environmental benefits. Nor has there been a clear link between such forms and economic viability. The acceptability of these solutions is untested, and the extent to which they might change behaviour towards more sustainable lifestyles is simply not known (CityForm 2008).

So how justified are these notions and wide-ranging claims of sustainability? Is there room for complacency? In the United Kingdom, Europe and much of the Western world, these solutions provide a comforting model, and although untested, they are being built. Yet there has been little fundamental research to assess the extent to which particular urban forms or physical design affects sustainability. Questions about its validity remain. For every positive claim there are also negative impacts. Higher densities may lead to overcrowding and more traffic, and may not be the choice favoured by residents wishing to purchase homes – population trends still tend to favour rural or suburban locations. The social and economic impacts are also hard to trace to the influence of urban form, as many other factors, such as poverty or employment, are more important.

If we cannot be sure of this in the context of Western cities whose growth is relatively slow and whose development is mature, what then of the fast-growing and very large cities in the rest of the world? It seems that in these cases theory, which is largely Eurocentric, may have exceeded its capacity. Theory simply cannot be scaled up to fit a context of fast-growing megacities and megacity regions.

BIG CITIES, BIGGER PROBLEMS?

Yet surprisingly, compact-city ideas have resonated beyond the Western world and have been taken up by countries where the urban context is very different (Jenks and Burgess 2000). Nowadays the most obvious point to make is that over half the world's population is urban. Across the world there are some 60 metropolitan regions with populations of more than five million inhabitants. Of these regions, nearly half have populations exceeding ten million, the largest being the Tokyo Metropolitan Region with more than 36 million people. About two-thirds of these large regions are to be found in Asia, where the growth rate is fast (World Gazetteer 2005; United Nations 2006). These vast areas, with very large cities at their core, suggest that the compact city and similar ideas are simply a contradiction in terms.

The criteria that, in the West, define sustainable urban form are also found in the world's largest and fastest-growing cities – and what is more they are found in abundance or, in many cases, over-abundance. Table 5.1 makes a simple point and, although the indicators are non-scientific, suggests from observation how such cities in different regions of the world might be ranked. Both cases fulfil the criteria in one way or another, but their application has little meaning, as cities such as Bangkok, for example, are hardly sustainable. However compelling the arguments may seem, however attractive the models are (like Amsterdam and Barcelona), as far as urban form is concerned, the criteria are not easily transferable. Those two words, compact city, that conjure up such beguiling images, need to be treated with caution.

The nostrum that a solution is to be found in high-density development also is problematic. What is high density? High density in London has a

Table 5.1 Criteria for the sustainable city – same difference?

	Compact city; new urbanism	*Cities in developing countries, e.g. Bangkok*
Mixed use	•	•••
High density	•	••••
Street grids	••	••
Socio-economic diversity	••	•••
Variety of transport choices	••	•••
Walkable neighbourhoods	•••	•

Table 5.2 Comparative urban densities

	Inner city	*Metro region*
Hong Kong	7.5	10
Cairo	4.5	8
Bangkok	2	2
New York	1.5	1
London	1 (7,250 people/km^2)	1 (600 people/km^2)
Los Angeles	−2.5	−3.5

completely different meaning, and exists in different urban forms, from that in Hong Kong (see Table 5.2). The naivety of assuming that what might work in one context and culture will work in another is demonstrated in the much-heralded 'world's first sustainable city', Dongtan, within the Shanghai metropolitan region (ARUP 2008; Cai 2008). Dongtan is being aggressively promoted as a model for China, and yet it is clear that as a model it is little different in form and density, and other technicalities, from schemes such as Greenwich Millennium Village in London. For China and Shanghai, for all its technical wizardry, it looks likely to become a relatively low-density, rich, upper-middle-class suburb for commuters to Shanghai – anything but socially and economically sustainable. Indeed, Hong Kong may provide a better model with its multiple and intensive land use (Lau *et al*. 2005; Figure 5.1a and b, colour insert 1)

Understanding the dimensions of the sustainable city is a complex issue. Care needs to be exercised over the context within which cites exist, their cultural background and regional and national differences. There will be significant differences in different parts of the world, but even so there are some common underlying and enduring themes that appear to inform both the debate about urban forms and sustainability – in particular the forms that are evolving in the largest cities and in metropolitan regions.

POLYCENTRISM

As metropolitan regions and cities grow, they tend to move from traditional monocentric forms towards polycentric structures. This is a phenomenon that occurs at different scales of the built environment, whether at the regional or city level, variously termed polycentric regions, polycentric urban systems, multimodal urban systems (centres, subcentres or nodes), or networks, urban networks or polycentric networks (linkages between sub-centres or nodes). Polycentrism is a form that either has evolved over time, or results from globalization and urban growth, or it is an objective planner's wish to achieve through policy (Jenks *et al*. 2008).

Where polycentric forms just happen, as in many North American cities, such as Los Angeles, and in cities in developing countries, they are often characterized by themed, gated communities, shopping malls, business parks and outer-city slums, mainly connected, but also separated, by highways. By contrast, the European ideal, for example the Randstad in The Netherlands, demonstrates policy objectives to achieve a controlled polycentric metropolitan region that is well-connected, with high-speed rail networks making the major cities of the Randstad (Amsterdam, The Hague, Rotterdam and Utrecht) accessible within one hour's travel (Ministry of Housing, Spatial Planning and the Environment [VROM] 2001).

As cities expand, particularly through rapid growth, the emerging new centres will form the nodes of a more polycentric urban form. If such nodes followed compact-city concepts, managed their transport and energy efficiently and encouraged walking or cycling, they would at least have stronger claims to sustainability than if growth were unrestrained. However, polycentric nodes surrounded by a network mainly connected by highways or where the social contrasts are stark can hardly be considered a sustainable urban form.

The change in urban form from monocentric to polycentric is in effect a breaking up of the city into new centralities, often with an autonomy or life of their own. While linking these centralities with good transportation may attempt to make up a whole, more often polycentric forms intensify rather than reverse fragmentation.

FRAGMENTATION

The inexorable growth of urban areas, the increase in the size of cities and urban agglomerations extending into vast regions means that problems of urban fragmentation have increased. The planned polycentric regions in Europe try hard to ensure there is an efficient interconnection between them with an efficient public-transport infrastructure, but it is a struggle to find sufficient funds for the comprehensive coverage needed. Fragmentation still exists with spatial disparities between rich and poor, and dependency on roads and private transport. In the megacities there are new centres that are large enough to be semi-autonomous and are effectively mini-cities in themselves, and that are often poorly interconnected. In the cities of the developing world there is widespread poverty and a huge disparity between rich and poor, spatially characterized by informal settlements and tenuous development on the urban peripheries (Kozak 2008).

There are many obvious signs of urban fragmentation, the clearest being the increasing numbers of gated communities – defensive bastions against the poor, or selling 'desirable' lifestyle aspirations to the middle classes. These exclude the majority, and often exclude real life and the benefits of the urban experience. The privatization of housing areas is similarly reflected in many city centres or in new centres through the privatization of space – the ubiquitous shopping mall being a prime example (Figure 5.2, colour insert 1).

The wealthy also contribute to the pressures to fragment in other ways, by the rich, the middle class and speculators capturing resources, particularly housing. This can result in land freezing, deliberately keeping property empty to take advantage of land price rises, and again, excluding the disadvantaged to peripheries or the run-down areas in the cities (Fahmi 2008).

Even the best of intentions can lead to urban fragmentation. Urban regeneration is a much-heralded policy for bringing life and wealth back to city centres. This has undoubted benefits in improving the environment of often run-down urban areas. In many cases the regeneration concentrates on the public realm, providing open space and areas to be enjoyed by all citizens and visitors (Figure 5.3a and b, colour insert 1). However, this also too often leads to gentrification, steep increases in land and property values and the banishment of the poor and even the middle classes to the peripheries.

The recognized need to link these disconnected places with public transport can also end up spawning another form of fragmentation. Unless it is

Figure 5.4a Disparities in transport and use for the poor and the well-off.

Figure 5.4b.

very heavily subsidized, which is unlikely in a world dominated by neoliberal economies, mass rapid transit provides an efficient but expensive mode of transport and thus becomes the preserve of the well-off (Figure 5.4a and b). Roads dominate transport; for the very rich through choice and for the poor through lack of choice. In many developing countries poorly maintained but cheap buses, motor cycles and cars are the dominant mode of transport, on usually inadequate and congested roads. Again, the rich benefit from toll roads and urban motorways, while the less well-off suffer the public roads. The roads, especially the urban motorways, add to the physical fragmentation of the environment (Charoentrakulpeeti and Zimmermann 2008).

The above analysis does not address the meta-level story about the impacts of globalization, international investment and the spatial side-effects of those who are excluded from the claimed benefits of globalization.

THE DEFRAGMENTED CITY

What can be done to defragment the fractured cities? Polycentric forms and their tendency to lead to urban fragmentation pose new problems in achieving a measure of sustainability. A new paradigm is needed that may incorporate relevant aspects of compact-city theory, but that is meaningful not just physically, but also culturally, especially in a non-Western context where most urban growth occurs. This may require a process of integration, of bringing together urban, social and economic fragments – of defragmenting the city.

When leading the Urban Task Force (1999), Richard Rogers claimed that UK urban design would provide the model for others to follow to achieve the sustainable city. Far from it. We fool ourselves if we think that planners, urban designers and architects can solve all these problems, but the built environment professionals could provide a framework within which defragmentation can take place. In some aspects (see Table 5.3) there is sufficient evidence to feel confident that they contribute to sustainability, but beware – these aspects must respect local culture and form.

Table 5.3 has significant omissions – the physical environment is just one aspect. Social sustainability – equity, social justice, poverty and social exclusion – all need to be tackled. To achieve economic sustainability, income inequalities need to be addressed, and employment, education and training, local business, services and facilities need to be encouraged. Lifestyles and attitudes need to change; ever more consumption, with higher expectations, driven by a celebrity culture seems ultimately unsustainable.

The traditional view of the compact city needs to be seen from a different perspective. A network of compact centralities or cities may come into being. The key is not just the nodes, but the network links between them that are equally significant (Figure 5.5). These transport networks must not priced so that they exclude the poor – indeed they must be affordable to all.

This can never happen unless there is significant public investment in the public good. Neoliberal economics favouring private investment for the private good is doomed to failure if sustainability is to be the goal. The excuse that China is now responsible for much the same level of greenhouse-gas

Table 5.3 Aspects of a sustainable built environment

Land use and built form	Environmental energy conservation	Environmental recycling and reuse	Communication and transport
Intensive use of urban land	Combined heat and power (CHP) – local power generation	'Grey' water systems	Light-transit routes, eco-friendly buses and bikeways
Networks of green corridors	Micro-power generation	Recycle water for gardening and car washing	Car clubs and cycle facilities
Community buildings, self-managed	Renewable energy	Reuse water and filter, to be directed to ecology parks or green spaces	Pedestrian-friendly infrastructure
Mixture of uses at relatively high density	Reduced energy consumption and embodied energy	Waste recycling, and use for production of biogas	Restricted car parking
Affordable homes	High levels of insulation	Reduced domestic and construction waste	Environmental advice – bus/transit times, energy and water monitoring
Local identity	Intelligent lighting and integrated security, heating and IT systems	Carbon-neutral lifestyle	IT enabled
Sustainable building materials	'A'-rated white goods Eco-rating, e.g.		
Flexible design and good space standards	BRE Environment Assessment Method, 'excellent'		
Improved noise insulation			

Sources: inter alia, CABE, 2008; English Partnerships, 2007.

emissions as the United States is risible, when a Chinese citizen per capita consumes around one-sixth of that of an American citizen. For the poorest countries, higher standards of living are justified, but for the wealthiest, reductions in the over-consumption of resources are needed. It may be that polycentric cities and regions can take forms that are more integrated – that a process of defragmentation can occur within a physical framework of compact nodes that enables citizens to behave more sustainably. The defragmented city must be one that is not only physically defragmented, but also socially and economically defragmented as well.

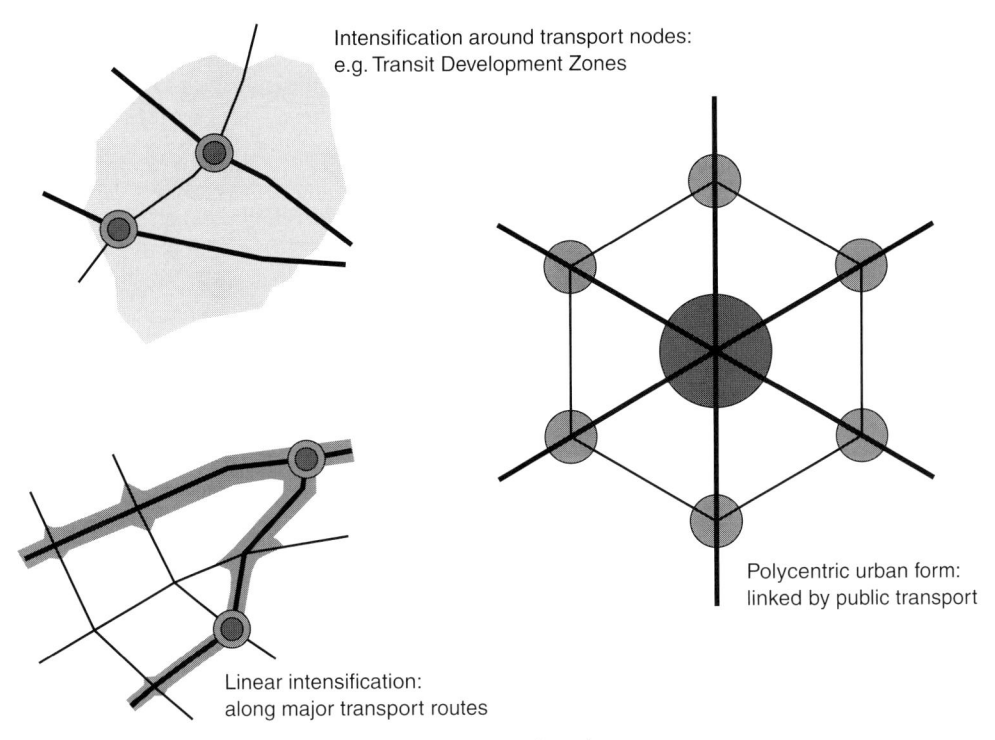

Figure 5.5 A network of compact centralities or cities (source: Mike Jenks).

NOTE

1 This chapter draws on many sources, including two books: M. Jenks, D. Kozak and P. Takkanon (eds) (2008) *World Cities and Urban Form: Fragmented, Polycentric, Sustainable?*, Oxford: Routledge, and M. Jenks and C. Jones (forthcoming 2009) *Sustainable City Form*, The Netherlands: Springer. It also draws on a number of presentations, including evidence given to a Royal Commission in the United Kingdom, and a major research project led by the author, 'City Form: the Sustainable Urban Form Consortium', funded by the Engineering and Physical Sciences Research Council under its Sustainable Urban Environments programme (grant number GR/S20529/01).

BIBLIOGRAPHY

ARUP (2008) 'Dongtan Eco-city'. Available online: www.arup.com/eastasia/project.cfm?pageid=7047 (accessed 6 May 2008).

Cai, L. (2008) 'Establish Chongming as the first Ecological Demonstration District in China', Dongtan Development Co. Available online: www.dongtan.biz/english/zhdt/plan.php (accessed 6 May 2008).

Canmore Housing Association (2008) 'Edinburgh: Slateford Green Housing'. Available online: www.edinburgharchitecture.co.uk/slateford_green_housing.htm (accessed 6 May 2008).

Charoentrakulpeeti, W. and Zimmermann, W. (2008) 'Staunchly middle-class travel behaviour: Bangkok's struggle to achieve a successful transport system', in M. Jenks, D. Kozak and P. Takkanon (eds) *World Cities and*

Urban Form: Fragmented, Polycentric, Sustainable?, Oxford: Routledge, pp. 302–20.
CityForm (2008) Homepage. Available online: www.city-form.org (accessed 6 May 2008).
City Mayors (2007) 'The largest cities in the world by land area, population and density'. Available online: www.citymayors.com/statistics/largest-cities-density-125.html (accessed 6 May 2008).
Commission for Architecture and the Built Environment (CABE) (2005) *Better Neighbourhoods: Making Higher Densities Work*, London: CABE and the Corporation of London. Available online: www.cabe.org.uk/default.aspx?contentitemid=671 (accessed 6 May 2008).
Commission for Architecture and the Built Environment (CABE) (2008) Homepage. Available online: www.buildingforlife.org (accessed 6 May 2008).
Commission of the European Communities (1990) *Green Paper on the Urban Environment*, Brussels: European Commission.
Department for Communities and Local Government (DCLG) (2006a) *Planning Policy Statement 3: Housing*, London: Her Majesty's Stationery Office.
Department for Communities and Local Government (DCLG) (2006b) 'Sustainable communities: building for the future'. Available online: www.communities.gov.uk/publications/communities/sustainablecommunitiesbuilding (accessed 6 May 2008).
Department of the Environment, Transport and the Regions (DETR) (2000) *Our Towns and Cities: The Future – Delivering and Urban Renaissance*, Cm 4911 Urban White Paper, London: Her Majesty's Stationery Office.
Department of the Environment, Transport and the Regions (DETR) (2001) *Planning Policy Statement 13: Transport*, London: Her Majesty's Stationery Office.
English Partnerships (2007) *Millennium Communities Programme*, ENG0049, Warrington: English Partnership.
Fahmi, W. (2008) 'The right to the city: stakeholder perspectives of greater Cairo metropolitan communities', in M. Jenks, D. Kozak and P. Takkanon (eds) *World Cities and Urban Form: Fragmented, Polycentric, Sustainable?*, Oxford: Routledge, pp. 269–292.
Jenks, M. and Burgess, R. (eds) (2000) *Compact Cities: Sustainable Urban Forms for Developing Countries*, London: E & FN Spon.
Jenks, M., Burton, E. and Williams, K. (eds) (1996) *The Compact City: A Sustainable Urban Form?*, London: E & FN Spon.
Jenks, M., Kozak, D. and Takkanon, P. (eds) (2008) *World Cities and Urban Form: Fragmented, Polycentric, Sustainable?*, Oxford: Routledge.
Katz, P. (1994) *The New Urbanism: Toward an Architecture of Community*, New York: McGraw-Hill.
Kozak, D. (2008) 'Assessing urban fragmentation: the emergence of new typologies in central Buenos Aires', in M. Jenks, D. Kozak and P. Takkanon (eds) *World Cities and Urban Form: Fragmented, Polycentric, Sustainable?*, Oxford: Routledge, pp. 239–258.
Lau, S., Wang, J., Giridharan, R. and Ganesan, S. (2005) 'High-density, high-rise and multiple and intensive land use in Hong Kong: a future city form for the new millennium', in M. Jenks and N. Dempsey (eds) *Future Forms and Design for Sustainable Cities*, Oxford: Architectural Press.
Ministry of Housing, Spatial Planning and the Environment (VROM) (2001)

Summary: Making Space, Sharing Space: Fifth National Policy Document on Spatial Planning 2000/2020, National Spatial Planning Agency, The Hague: VROM.

Smart Growth (2008) Homepage. Available online: www.smartgrowth.org (accessed 6 May 2008).

Stubbs, M. and Walters, S. (2001) 'Car parking in residential development: assessing the viability of design and sustainability in parking policy and layout', in RICS Foundation, *The Cutting Edge 2001*, London: Royal Institute of Chartered Surveyors (RICS).

United Nations (2006) *World Urbanization Prospects: The 2005 Revision*, United Nations, DESA, Population Division. Available online: www.un.org/esa/population/publications/WUP2005/2005wup.htm (accessed 6 May 2008).

Urban Task Force (1999) *Towards an Urban Renaissance*, London: E & FN Spon.

Williams, K., Burton, E. and Jenks, M. (eds) (2000) *Achieving Sustainable Urban Form*, London: E & FN Spon.

World Gazetteer (2005) Homepage. Available online: www.world-gazetteer.com (accessed 6 May 2008).

Part II

OTHER CULTURES, APPROACHES AND STRATEGIES

INTRODUCTION

Jenks' proposition about the need for different paradigms brings us to the second group of chapters. This cluster presents examples, design initiatives and thinking from a number of cities in countries that are both geographically and culturally remote from those in the West – Japan, Thailand, Singapore, China and India. Here we encounter cities that shrink and those that uncontrollably expand and cultures that traditionally value things that are fundamentally different from those glorified by the (stereo)typical West. How are we to think about the sustainable future of those cultures and cities?

Ohno Hidetoshi opens this cluster of chapters with a discussion of his recent urban project, Fibercity 2050. He focuses on the problem of shrinking cities, which is beginning to emerge in Japan, but is also relevant in other parts of the world that are experiencing a decline in population numbers, either due to ageing or voluntary or forced migrations. For shrinking Tokyo, Ohno proposes urban fibre as an alternative metaphor for understanding the metropolis. In a way that is sensitive and responsive to culture and to the precise demands of place – both as existential space and as time – Ohno proposes four strategies to realize this concept. They aim to transform the character of the city by carefully manipulating its existing linear elements, or fibres. The idea is to deal simultaneously with large and small scales and with long and short urban rhythms, while never losing sight of the human scale, which is one of core values behind the idea of eco-urbanity.

Ishikawa Mikiko offers another view of Japanese urban spaces, their current problems and the possibility of making incremental improvements to them. She uses as an example her project in Kakamigahara City to explore the theme of the lost commons, which are an important part of eco-urban infrastructure. Historically, Ishikawa reminds us, the commons were places where local ecological and cultural spheres interacted. From the 1960s Japan's rapid urbanization led to the destruction of many of its commons, especially those located on the urban fringe. Ishikawa uses the example of Kakamigahara City to describe the planning process needed in order to facilitate the social consensus that she sees as being both the test and the precondition of the success of any urban scheme. She concludes that the success of Kakamigahara and other regeneration projects depended on the combined efforts of many, and in particular the involvement of individuals who might not speak out eloquently, but who work diligently for their communities. Her chapter thus reinforces the essential importance of community participation in eco-urbanity if we are to make and live in sustainable cities.

Using the case of Singapore's Nankin Street, Heng Chye Kiang introduces another approach to urban regeneration and discusses the dialectics of imported (or imposed) practices and those that stem from local conditions and culture. The city of Singapore has a short history of only 188 years, of which the first 140 were under British rule, followed by Japanese occupation in the mid-1940s. With the formation of the first non-colonial government in 1958, the colonial footprint in Singapore began to be modified by different agendas and priorities. A new democratic mandate generated new spaces and activities in the oldest parts of the city that built upon and modified the colonial layer, rather than obliterating it. Heng's contribution examines and discusses the transformation of areas in the heart of the city of Singapore during the last 30 years. As his analysis shows, such new urban layers can take the form of changes to the physical grain and texture of the city and its spatial extent, while also allowing whole new functional roles to emerge in the existing fabric. Led by different agencies, the transformation engendered radically new urban tissue along certain stretches, while conserving and adaptively reusing the historical fabric along others. Both the new departures and physical and cultural continuity are present along the now pedestrianized Nankin Street, which itself is part of a larger network of public space that has taken shape over the last two decades, during which new as well as familiar spatial and cultural activities have evolved.

Sidh Sintusingha addresses the issue of urban sprawl through an examination of two geographically and culturally distant places: Bangkok and Melbourne. Sintusingha sees urban sprawl as a phenomenon where globalization overlaps with unsustainability and local cultures that take the form of pre-existing settlements, rural patterns and cultural perceptions of development and nature. He reminds us that urban sprawl poses some of the most critical challenges for urban sustainability today, and that this common global phenomenon is closely tied to economic trends and population growth. His chapter suggests that local factors need to be investigated, and he discusses the possible synthesis and reconciliation of the generic notion of sustainable cities (as spatial and urban forms), specific cultural landscapes of production and reproduction and evolutionary trajectories (including socioeconomic and political environments). Sintusingha hopes that they could form a basis for policies, planning and design towards sustainability, in both quality and diversity.

Jianfei Zhu focuses on the current busy pace of Beijing, and the possibility and necessity of learning from its local past. His starting point is that, after years of robust economic development, public space and the public interest in China have been neglected. How, though, can we reconstruct the public good in Chinese cities, in both social and spatial terms? Zhu employs Henry Lefebvre's distinction between the formal, frontal space of rationality (representation of space) and the informal space of irrationality (representational space) to revisit traditional Beijing and its urban patterns and offer the 'socially ecological' qualities of the Chinese urban tradition as a remedy. Lefebvre distinguished between frontal rational space and a dark space of desire and informality found in post-Renaissance Europe. He suggested that Asia and China, due to the characteristics of Chinese writing where the abstract and the figural are not separated, do not employ the distinction found in Europe. Zhu argues that in traditional Beijing we find urban patterns where there is no distinction between frontal space and the messy space of informality and irrationality that is neither abstract nor figural. He hopes that new conceptual tools may be developed from this point that can help to reconstruct an appropriate space for people in cities in China today.

In the chapter that concludes the second cluster of themes, Arvind Krishan points out that, despite an increased appreciation of the desirability of eco-urbanity, in practice massive compromises are made with it, mainly for economic reasons. Krishan reminds us of the social and ethical impacts of globalization, climate change and income disparities between the rich and the poor. He emphasizes that the development of human settlements has a critical role in correcting some of those wrongs. This chapter seeks to set stage for dialogue on practical, achievable actions for developing sustainable settlements. He examines what constitutes an eco-settlement and proposes a range of the imperatives that different stakeholders need to meet if new settlements, as far as possible based on an eco-logic, are to succeed. He puts forward two kinds of arguments, one related to quality of life or liveability in a culturally appropriate settlement, while the other relates to the way the settlement functions. The chapter also leads us away from large, urban-scale projects and initiatives down to architectural scales and Krishan's own projects in some of the most environmentally and culturally vulnerable parts of the world.

6 DESIGNING FOR SHRINKAGE
Fibercity 2050, Tokyo

OHNO HIDETOSHI

SHRINKAGE CANNOT BE AVOIDED

If the twentieth century can be characterized by growth or expansion, the greatest issue for the world in the twenty-first century is shrinkage, which can take various forms and results from a variety of causes. First, there are global environmental concerns. The earth no longer has the capacity to absorb or handle emissions from the voracious productive activities of humankind. Limiting CO_2 emissions is an urgent issue, so the twenty-first century will experience great restrictions on production. Furthermore, natural resources will not suffice if people all over the world consume resources at the same level as advanced nations do today. Meanwhile, the world's population continues to expand. Due to finite natural resources, the amount of resources available per person will decrease. While the speed of technical improvements will contend with these limits to some extent, if human desires continue to escalate, there will be no way to avoid shrinkage. Although the world's population is increasing, the birth rate in one-third of all nations is already so low they cannot maintain their populations.

The contrary phenomenon, lower birth rates combined with an ageing population, is occurring in many advanced nations. In 2005 in Japan the population reached its peak size, after which a long-term period of population reduction began. By 2050 it will have decreased to three-quarters or less than the present population, and senior citizens (those aged 65 years or older) will constitute 40 per cent of the entire population. In all areas, as affluence improves the birth rate falls and longevity increases. In the twenty-first century many regions worldwide will face shrinkage as a result of both population decrease and environmental problems, and political tensions will result. If this shrinkage continues at its present rate, societies and urban environments will be devastated. As urban designers and architects, we need to acknowledge the challenge posed by shrinkage and work to acquire the knowledge that will change this misfortune into good fortune. Shrinking resources and population size will have a large, direct impact on the city. Up until now, however, no urban design proposals have been premised on the shrinkage of society. Because human society worldwide has been continually expanding, whether slowly or rapidly, since the beginning of recorded history, it is hard to imagine, let alone plan for, the opposite phenomenon. Fibercity is one attempt to do so, and furthermore to treat shrinkage as an opportunity for urban improvement.

WHY CONSIDER THE CITY IN TERMS OF LINE?

Likening urban planning to an organ transplant, problematic areas of the city require buildings to be demolished and replaced with completely new structures, for which surface interventions cannot be avoided. In mature democratic countries, however, save for the eventuality of a natural disaster or war, attempts to alter the surface of existing urban areas take a long time to realize. But, although architects claim that architecture can change a city, architecture consists of interventions at discrete and separate points, not over whole areas, and it is unlikely that this on its own will change the city.

The alternative proposed here, Fibercity, focuses on the line that lies between the surface and the point. In geometrical terms a place cannot be defined only by points, but we can define an area with a line. Moreover, if we compare long, linear parks with a square one covering the same area, we see that the former has a greater perimeter and thus creates a greater amount of desirable housing lots alongside it. Currently redevelopments in contemporary Japanese cities merely intervene at specific points, so that the architectural facilities that result can only be seen from their immediate environs, or, like a theme park, the redeveloped areas are surface interventions enclosed like islands in the existing cityscape. In either instance, the approach seeking a slow transformation of the city through involving the surrounding areas is lacking.

With limited resources and a need to value and pass on the embedded memories that have been carved into the city, we should approach twenty-first-century urban design with the idea of making linear interventions into existing urban areas with the expectation of a resulting gradual transformation of their neighbouring areas. This approach to urban design is like making cloth by weaving fine fibres that mutually intertwine without the use of glue. Thus, our proposal for a city that is comprised of the fibres of rivers, cliffs, tree-lined streets, railways, shopping avenues, riverside developments, bridges and long, collective housing complexes is entitled Fibercity (Figure 6.1, colour insert 2).

THE FIBERCITY AS A COUNTER-MODERNIST PROPOSAL

While many cities developed around trading and trade routes, the modern cities that emerged from industrial capitalism were based on production. For this reason, modern urban planning did not pay much attention to the concept of exchange (of such things as commerce and information). Instead of attempting to relate zones of different usage to each other, the emphasis was on discontinuity and on creating regions with a limited range of activities. However, contemporary cities are again underpinned by the exchange of goods and information. When the focus shifts to exchange, it is directed towards the boundaries between areas rather than towards the area itself.

When, prior to the modern era, cities were established on the basis of exchange, most of their inhabitants' daily lives were expressed and contained within the city. In the contemporary city, people and goods move about beyond urban boundaries, and cities exist in a state of mutual dependence. Speed is what supports this interdependence. In the contemporary city, high-speed movement is becoming more and more important for industry as well as for the individual. Although the charm of a large city lies in the amount of options available there, this, too, is made possible by popular high-speed transport. Thus, it is more suitable to use lines than surfaces to control the contemporary city.

Modern urban planning developed upon the premise of population increase during the Industrial Revolution, and this way of planning has spread worldwide against the backdrop of the post-Second World War population explosion. New creations, in other words inventions, therefore merited praise. In the cities of modern post-industrialized nations facing a population decrease, however, a material shortage is becoming much less of a problem. Instead of requiring technologies of invention, therefore, ways to optimize the existing environment should be sought. This can be conceptualized as editing. Although in modern urban planning the idealized city has been conceived as an elaborate machine, a flexible fabric might be a more appropriate ideal for the twenty-first-century city or, in other words, a Fibercity.

THE GREEN FINGER

A reorganization strategy for suburban areas affected by population decline

The notion of a Green Finger is a strategy for reorganizing suburban areas that are severely affected by population decline (see Figure 6.2, colour insert 2). The strategy proposes gradually concentrating vast stretches of suburban housing around zones within walking distance of stations along railway lines and the greening of outlying areas beyond, transforming them into desirable patterns for residential use. Suburban residential areas are a characteristic form of modern city organization. These specialized residential areas were designed for nuclear families living in detached individual houses managed by full-time housewives, from which one had to travel to the city centre in order to work or consume. But in an ageing society with a low birth rate, the rationale for the suburbs fails. First, single-person households will become the most dominant household unit. Moreover, due to the population decline of the generation who formerly supported them, women and senior citizens who are in good health will have to work, thus making the commute to your place of work a major criterion in selecting your area of residence. Further, due to the limitations in physical strength and economic power of the elderly, who will constitute over a third of the population, the ratio of homes with a private car will decrease, making car-dependent areas undesirable places to live. Since householders who own property are likely to move promptly to places that are more accessible near the city centre, only the poor households will be left behind in the remote suburbs if no measures are taken.

Along with population decline, it is likely that the overcrowding in large Japanese cities will be resolved. Although from a statistical point of view the residential land area per person will increase, housing lots are already being

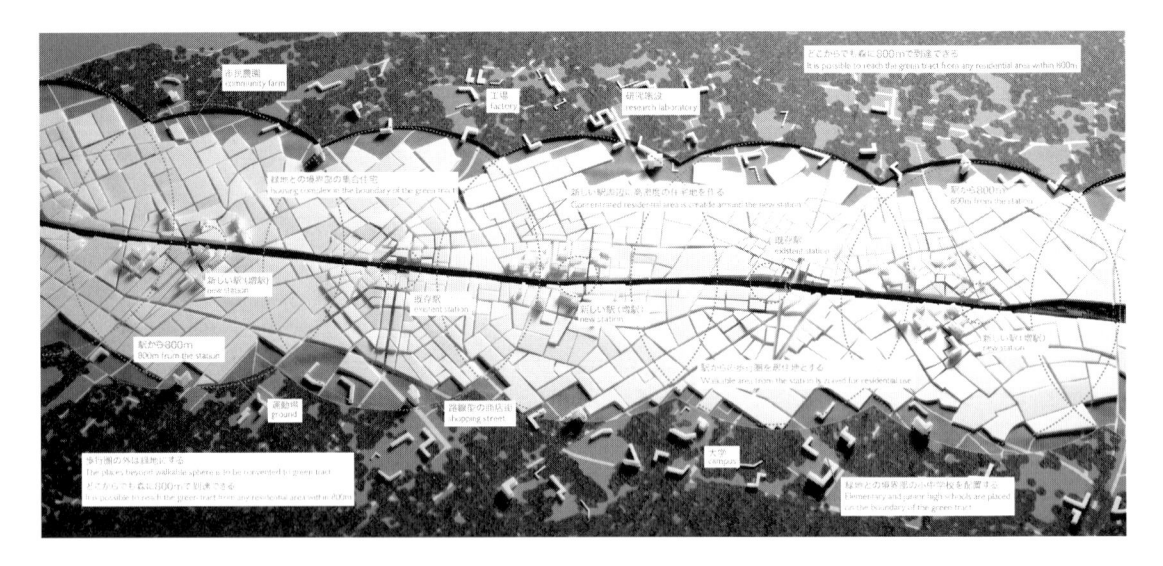

Figure 6.3 Compact Town and Green in the Green Finger.

subdivided into small parcels. You cannot expect a house in the right area to come up for sale at the time you want to buy it. Thus, most housing lots will continue to remain the same size and there will be an increase in vacant properties scattered through the suburbs.

The Green Finger strategy proposes that residential areas be established within walking range of train stations, and their surroundings be revived with green areas. Business facilities such as universities and research laboratories can be asked to move to these green areas and to offer part of them up to the public as rentable kitchen gardens or farmland, all the while maintaining a high ratio of greenery. While housing lots will remain small in size, the convenience of the city will be ensured, since there will always be a train station within walking distance, and if you walk in the opposite direction, there will be a vast expanse of green area that does not exist in today's metropolitan area.

The 294 compact cities

There is global recognition that dispersed, car-dependent suburban residential areas are inappropriate for contemporary environmental problems and an ageing society. Against this background, it is the compact city that has been recently attracting interest as a sustainable urban form. This is an urban form that does not depend on the motor car and in which people live only within walking distance to their daily needs and activities. However, while the compact city is an environmentally rational concept, doubt remains over whether it would win the support of many people. This is because there is a worldwide tendency for younger people to reject smaller cities and be drawn only to large ones. One aspect of this attraction to metropolitan areas is the abundance of choice and opportunities they have to offer. Because everything is on offer, from sources of employment to educational institutions, culture and entertainment, consumerism, encounters with the opposite sex and red-light districts, the metropolis displays altogether richer possibilities than a small city.

This abundance of choices is made possible by transportation, and Japan's metropolitan areas have well-developed, world-class railway networks. Whether in terms of route density, frequency of trains that arrive on time or safety, this network leads the world. Yet as the population density in the suburbs decreases, it will become economically difficult to maintain the suburban railway lines. If these lines are abandoned for economic reasons, the poorer residents of the more remote suburbs (including many senior-citizen households) will be left stranded. By setting up residential areas within walking range of train stations the Green Finger will make better use of our well-developed metropolitan railway network. It aims towards an urban form in which, despite population decreases, the railway infrastructure can be maintained and automobile dependency decreased, contributing to the reduction of CO_2 emissions. There will be 730 compact cities around 730 stations linked to each other by the railway system. By stringing residential areas within walking distance of the train station along the railway, the Green Finger can achieve an urban form with both a compact-city environment and the allure of a large city.

THE GREEN PARTITION

A disaster prevention strategy for dense urban areas

Despite the recent improvements in the accuracy of earthquake forecasting and concrete disaster probability investigations, earthquake disaster prevention measures at the city-planning level have not progressed. The Green Partition is a strategy for arresting the spread of fire during earthquakes by dividing up and separating high-risk disaster areas – urban areas with a high density of wooden structures, known as the Mokuchin Belt, that extend around and enclose the city – into small sections with long and narrow green belts (see Figure 6.4, colour insert 2). Up until now there have been two types of planning improvements in these areas: widening roads and consolidating narrow housing lots to create fire-resistant collective housing complexes. The widening of the roads attempts to create firebreaks to prevent fire spreading, to improve access for fire vehicles and to secure escape routes for the inhabitants. But, in fact, not only does it take an enormous amount of time to buy up property from residents along the planned roads, the project cost is swelling daily as the cost of buying up the land increases. Furthermore, turning the small lots into collective or cooperative housing complexes is not welcomed by the residents. Thus, even if the results are successful, unless they are supported by the residents the project will not progress and these areas will unfortunately become more crowded during the delay to the project.

In contrast, the Green Partition proposal does not predetermine a desired outcome, but aims to connect up vacant lots as they emerge on the market for sale and to exchange land when necessary. The final form that the green partition will take cannot therefore be predicted as it will depend on the availability of vacant lots. However, in each case, one end will always be connected to an evacuation area, securing a safe passage. Moreover, due to the improved quality of the environment from the increased ratio of green coverage, the increased land price will balance out the reduction in housing lots.

Thus, in comparison with the conventional planning method in which people may be forced to accept a predetermined plan, flexible execution is necessary for realizing the Green Partition. Similarly, due to the sluggish way in which the municipality improves the built environment by using taxes, the enactment of this project is better suited to a non-governmental organization such as an urban planning corporation and licensed for this purpose by the municipality. More specifically, a method of allowing landowners in the area to become stockholders so as to receive assistance from the municipality should be considered, where the surplus floor-to-area ratio can be transferred to development sites in the city centre and managed by earnings from the sale of the property.

Green belts to prevent fire spreading

The objective of the Green Partition is to improve methods of disaster prevention while simultaneously improving the liveability of dense urban areas. For this purpose, a minimum width of 4 m is required for the linear green areas. According to research on fire prevention and the distance of fire spread, if a 4 m belt of unoccupied land is maintained between buildings, fire spread can be prevented even for a 25 m-wide building.[1] Moreover, for residential environments, a width of 4 m is required to ensure adequate ventilation, daylight and privacy, and 4 m is also the minimum width for roads that provide direct access to a building lot, in accordance with the Building Standard law.

By installing treed firebreaks in the green areas, the Green Partition can further improve the environment and the efficacy of deterring the spread of fire. The effectiveness of the broad-leaved evergreen trees which constitute the natural vegetation in regions south of Sendai, such as the thick, shiny-leaved evergreen native beech, *shii* (*Castanopsis*) and camellias, for preventing the spread of fire has been substantiated by the Hanshin-Awaji earthquake. However, they tend to create an undesirable amount of shade and so substitutes may be sought. Similarly, as green areas in residential districts, a comprehensive design for these partitions must include considerations of crime prevention, aesthetics and the usage of these green areas, as well as how they will be maintained and by whom.

The formation of the Green Partition begins with the selection of open spaces such as schoolyards to which the partitions will be linked. If there is a wide arterial road in the surrounding area, they will also be connected to it. The Green Partition does not only serve as a firebreak, it also allows the residents of dense urban areas to use it during fire evacuation to a vacant lot or arterial roads. It can also be used by fire vehicles and rescue operations. After selecting the pivotal vacant lots when a suitable housing lot becomes vacant, the town planning committee will requisition the land (either by purchase or lease) and green it. Housing lots that do not meet the requirements for the direct access road, narrow housing lots, and others that are poorly suited for residential use will take priority in being requisitioned. While it may be rational to connect destinations with the shortest distances for general roads, since

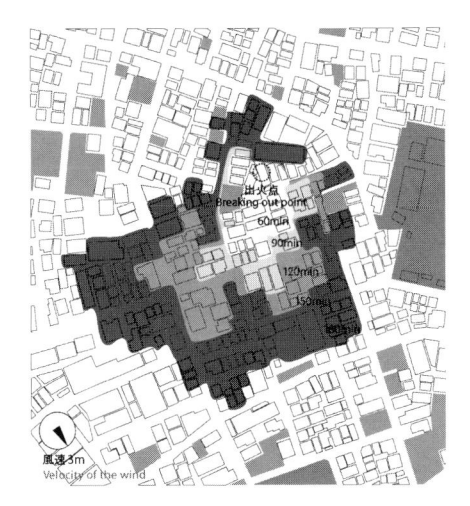

Figure 6.5a The fire-spreading process simulation in present condition.

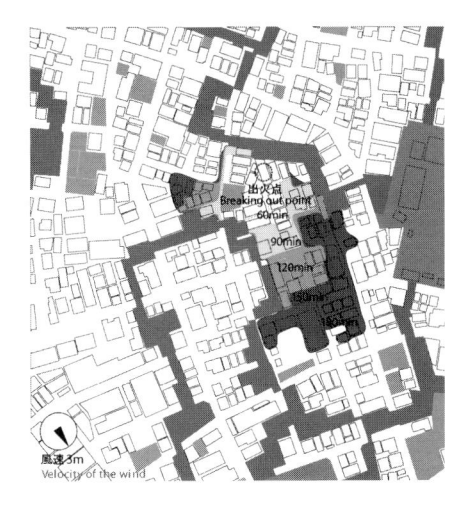

Figure 6.5b The fire-spreading process simulation after the linear green land completes.

the Green Partition winds about, it serves the intended purpose by linking a large number of districts.

THE GREEN WEB

Transforming the inner-city Tokyo Metropolitan Expressway into a green emergency road

The Green Web is a strategy to transform the inner ring road of the Tokyo Metropolitan Expressway into green roads and roads for emergency disaster relief. It will also enhance the use of the property along the roads and promote the introduction of regional energy systems (see Figure 6.6, colour insert 2). Should an earthquake – Tokyo's greatest threat – occur in the city centre during the daytime, the roads are likely to be buried in abandoned cars and collapsed buildings. The central functions of the state cannot be suspended for even a moment and a clear route would be needed for the vast number of evacuees to return once the earthquake is over. Thus, if one lane of the Tokyo Metropolitan Expressway that was reinforced after the Hanshin-Awaji (Kobe) earthquake can be secured for emergency vehicles, extremely prompt action can be taken.

Because plans for relocating the central functions of the capital to avoid such risks have been set aside, a back-up plan of this calibre seems to be vital for the capital, and there appears to be no other strategy that could be equally effective. It is proposed that the single car lane that is to be used as an emergency road should normally be used as a pedestrian road or cycle path. While the view of Tokyo scenery is haphazard and chaotic when seen from street level, the view from the Tokyo Metropolitan Expressway is quite beautiful. It is also proposed that the remaining lanes be greened. Continuous with the rooftop gardens on buildings along the route, this elevated green corridor will contribute to the enhancement of the environment of the city centre. As a linear park floating in the air, it is likely to become a worthy symbol of the innovative creation in Tokyo of a car-free society.

Although this proposal has been carefully conceived, since the Tokyo Metropolitan Expressway is a major artery for the vehicular traffic that supports 28 per cent of traffic in the city centre, it is likely that there are many doubts that it will ever be put in place. Yet when the central ring road currently under construction outside the Yamanote Line is completed, a traffic bypass through the Tokyo Metropolitan Expressway in the city centre is expected. Meanwhile there will be a decrease in the amount of motor-car traffic on the Tokyo Metropolitan Expressway corresponding to a population decrease of up to a quarter. These factors must be considered because there are a variety of methods being proposed to reduce vehicular traffic, such as the electronic road-pricing system. The reduction of greenhouse gases emitted by car exhaust is not only achievable through green engineering in the automotive industry, we must also decrease our dependence on cars.

The Tokyo Metropolitan Expressway as a city *meisho*

In preparation for the 1964 Tokyo Olympics, the Tokyo Metropolitan Expressway was constructed as part of an urban development plan utilizing the space above the rivers and roads without the need for expropriating land. This plan, which was then considered futuristic, is now seen as being responsible for the destruction of the cityscape. However, Tokyo's cityscape is composed of built-up layers of activities that have taken place in different ages. Just as the stone wall moats from the Edo period are considered part of the historical cityscape, so can the structure of the Tokyo Metropolitan Expressway be celebrated as a significant part of the historical cityscape of central Tokyo, albeit one in need of a new use. The concept of converting an elevated structure into a pedestrian road is not at all new. In both Paris and New York this type of conversion has either been actualized or is being considered.

The Tokyo Metropolitan Expressway would become a splendid riverside elevated pedestrian road alongside the Sumida River, overlooking the cherry blossoms along the bank (see Figure 6.7, colour insert 2). This would transform its hitherto gloomy image into a more positive and attractive one. At the point where it approaches the water surface in Chidorigafuji, a deck almost touching the water in the moat could be constructed for cyclists along the aerial road to take a rest. If part of its function as a public highway is moved to the underground tunnel of the Tokyo Metropolitan Expressway, the number of lanes above ground in the vicinity of Miyakezaka could be reduced to only one regional service road. In this way, the view of the most beautiful Imperial Palace moat in the city centre can be maintained and a pedestrian road can be provided. Greening the entire surface of the Tokyo Metropolitan Expressway will create a green area that is twice the size of the Shinjuku Imperial Gardens. Additionally, constructing roof gardens on buildings along the route at the same level as the Tokyo Metropolitan Expressway and connecting them to the Expressway with bridges will lead to a continuous horizontal extension of the green area that will heighten the charm and utility value of both (see Figures 6.8 and 6.9, colour insert 2).

The Tokyo Metropolitan Expressway as an energy centre

In order to solve environmental problems, we must decrease the enormous amount of energy being used by buildings. For this, while the greatest advantage of localized air-control systems is that energy use can be managed separately in each area to improve energy efficiency, in order to form energy networks our choices have been limited to areas of large-scale redevelopment. With the Green Web, an efficient next-generation localized air-control system can be achieved in existing urban areas by utilizing the underside of the elevated structure. Pipes can be installed above the roads or along the girders of the Tokyo Metropolitan Expressway, and by utilizing the space beneath the Expressway for the plant, the costs of both piping and plant construction can be greatly reduced. Furthermore, energy can be supplied from outside and thus cooling towers can be eliminated from the rooftops of buildings along the Expressway, further facilitating active greening of these rooftops. In this way, a green network incorporating the green areas of the Tokyo Metropolitan Expressway can be constructed.

It is necessary to combine fossil fuels and renewable energy as sources of energy. To raise thermal efficiency and apply ecological principles, a combination of such sources of co-generation, fuel cells, photovoltaic and wind-power generation and biomass should be used. In comparison to conventional individual air-conditioning systems, a reduction of roughly two-thirds in both energy and CO_2 emissions can be achieved by introducing this system. Moreover, construction costs for piping and plant building can be reduced to half

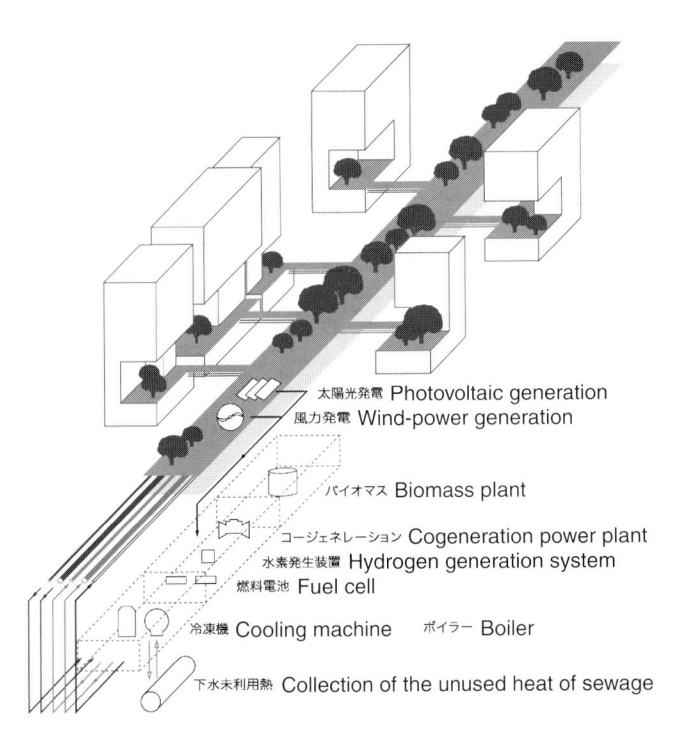

Figure 6.10 Localized energy system.

that of typical construction costs.[2] While it is not possible to overestimate the future demand for new buildings in central Tokyo, considering the demand for rebuilding, it is safe to assume that there will be a constant supply of new offices and housing. This being the case, we cannot entrust these to chaotic real-estate developments. Through the Green Web proposal, the large-scale conversion of housing lots into green roads along the Expressway as a priority will increase urban efficiency as it runs through the central districts. Hence, the prevention of excessive development in the central part of the distinctive nested spatial structure of the city block in central Tokyo is important.

THE URBAN WRINKLE

Creating new *meisho*

The Urban Wrinkle is the strategy through which distinctive linear spots of interest, or *meisho*, are created by the optimization of scenery and history of a locale within a homogenized and monotonous urban space.

Since Hiroshige Ando's paintings '*Meisho Edo hyakkei*' ('One Hundred Famous Views of Edo') of the former cities of Japan, well-known public spaces have been referred to as '*meisho*'. These were places where the history of the place and people's memories accumulated, places where one can relax and places of exchange. Unlike formal, artificial places such as the plazas of Western

Figure 6.11 Map of the Urban Wrinkle Tokyo.

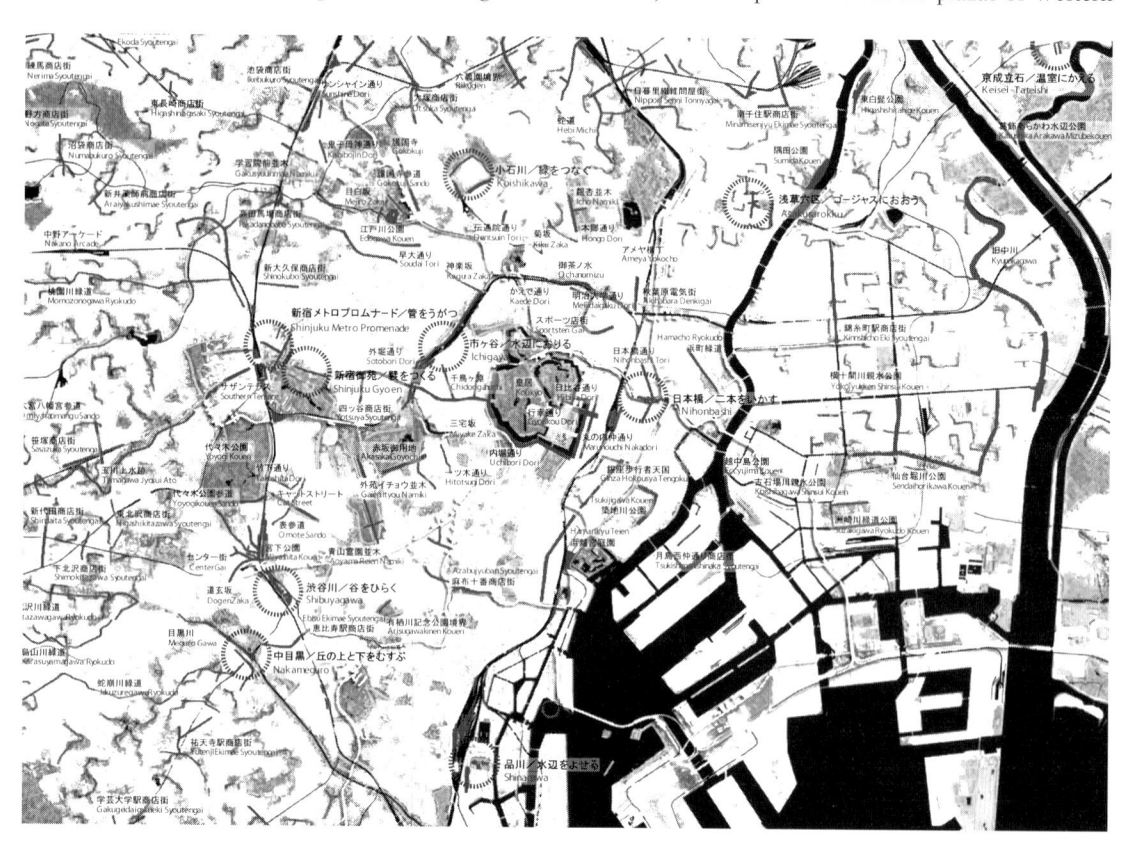

cities, *meisho* were places hidden in urban substrata that expressed fragments of nature and could even be referred to as interstices of the artificial order. Many of these disappeared, however, in the county's high-growth period (1956–73). In their place, plazas and streets emerged in theme-park-like 'consumption spaces' in the city. These spaces are produced skilfully, and as long as the *meisho* heritage exists, they satisfy the general public and the class-oriented vanity of city life. However, such pseudo-public spaces of consumerism will not survive in a depopulated society. Although many cities of Japan have managed to keep alive their pre-modern *meisho* inheritance, their expiry date is gradually approaching. It is time for us to create new *meisho* through our own efforts.

In Japanese urban spaces, *meisho* are often linear. For instance, Omotesando is a current *meisho* of Tokyo. It stands out from its surroundings as a well-defined area with a distinctive atmosphere. Although its shopping avenues are lined with the world's top designer-brand shops, it is a place that has particularity beyond international designer shopping. Here, the streets have become a special arena where passers-by stride in their finery with an air of affectation brought about by their lofty sensation of being on stage. Located in the residential area of Setagaya is another *meisho*, the Todoroki ravine. This deep ravine is carved out of the fluvial terrace of the Tama River by the Yazawa River, a small tributary. On either side of the ravine is a steep cliff with a thick growth of diverse vegetation that has a mountain-like charm not typically associated with residential areas. Both of these examples have a strong sense of place and are interstices of urban space that provide people with various options for passing the time.

Westernizing in the Meiji era and 'Project X' in the Showa era

At the beginning of 2006 an article in the newspapers reported the former Prime Minister Koizumi's call for a feasibility study to relocate the Tokyo Metropolitan Expressway that blocks out the sky above the Nihonbashi area. While we should be glad that we live in an age where politicians propose vast public investments for the sake of a scenic attraction, there is also a sense that this would be a waste of money merely to preserve second-rate historical remains, to say nothing of being patricidal urban design. Proposals that appear reformist often come in the form of reproaching your father's generation and nostalgically attaching yourself to that of your grandfather. In modern Japanese history, people in the Meiji era dismissed the Edo era, and in the post-war era, they rejected the pre-war days; now, they speak ill of the high-growth period. On the other hand, it appears that the dismal Japanese cityscape is due to the lack of value that the Japanese people place on their historical buildings. If we do not value the relics of our fathers' era, it is obvious that nothing will remain from our grandfathers' era.

It has long been said that the town context is important in a stocktaking era; yet, when faced with tangible projects, it is cheaper to destroy them than keep using them, and it is said that in Japan there is no streetscape worthy of respect. In our stocktaking era we must see that if we wish to imbue our towns with history, it is first necessary to be prepared to accept existing structures. The Tokyo Metropolitan Expressway must have been a former 'Project X'. In reviling it, we give our fathers' era a bad name.

In this proposal, the Expressway hovering over Nihonbashi Bridge will remain and will be suspended from a number of newly built arches that straddle the river, and the existing piers that support the elevated structure from the river bed will be removed. The surrounding buildings have increased to several times their height compared to when the bridge was first built, and, while the small Nihonbashi Bridge is buried between them, it displays a certain presence at this level. If the piers are removed, the sides of the Nihonbashi Bridge will be seen very clearly from the riverside as well. This proposal gives new life to the scenery of both the Nihonbashi Bridge and the Expressway, which were completed in different eras, instead of rejecting one of them (see Figure 6.12, colour insert 2).

Transforming the boundary of the Shinjuku Imperial Gardens

Fibercity is a design concept that focuses on the line. The boundary is a critical element as the form of a line in the city. It is based on the idea that urban design should return to the premise of exchange and interchange as the essence of the city. In cities designed by the pioneers of a modern town planning, the boundary was considered a device to separate clearly defined zones, such as residential, commercial or industrial areas. From a historical angle, however, cities began as places to exchange goods and information, and exchange is becoming increasingly significant in the contemporary city. Exchanges occur on the borderline of different areas. Typical boundaries are city gates, airports, ports or train stations. Retail facilities such as markets, shopping malls and department stores are all places of exchange. Residential districts for foreigners (like Chinatowns and Korea Towns), located on the fringes of the city, the back-alley night-life areas or military bases are types of boundaries. Places such as universities, cultural learning centres, language schools, libraries, museums, galleries and music theatres should also be boundaries for the exchange of knowledge.

Parks introduce nature into the city and also offer the inhabitants valuable opportunities for interaction. Most urban parks in Japan were based on the spatial composition of traditional Japanese gardens and are, as a result, closed to the outside. The shape of the site promotes this tendency. Although the Shinjuku Imperial Garden is a vast park, its presence is completely undetectable, even from within downtown Shinjuku. Lacking the urban presence that would enhance its charm, like Ueno Park and others, Shinjuku is a place to which one goes only on purpose. In accordance with the concepts of Fibercity, without moving the park area, the site boundary will be altered to a zigzag that lengthens its borderlines. In this way, while walking along Shinjuku Avenue, it will be possible to catch sudden glimpses of the vast green area and to be drawn to it for rest and repose. Since property facing the park would increase in land value, this change in the park boundary would also have a positive economic effect. Without any change in the house's floor area or project costs, landowners in such positions might be able to raise the value of their properties if proper steps are taken (see Figure 6.13, colour insert 2).

A proposal for a waterfront promenade along the edge of the Ichigaya moat

The reason that Tokyo has no charm as a city is because the facilities for vehicular traffic run rampant over the face of our city. At any train station, roundabouts are produced for cars, but bicycle parking spaces are shoved into a corner and treated as a nuisance. There are designated bicycle lanes in the towns of northern Europe, and they are even provided along the narrow roads in city centres. In Tokyo, the Metropolitan Expressway insensitively straddles valuable waterways and wooded areas. While Tokyo used to be a town like Venice, in which waterways wound through the streets, most have been paved over to become roads, and even to get near the precious remaining waterways, one must cross wide lanes of traffic. Walkways along the waterfront are narrow and their amenities are pitiful, where even taking a pleasant stroll is rarely possible. Although the government and capital attempt to attract foreign tourists to the city, what would draw them to a place with facilities for only commerce and transport?

Along part of the outer moat of the Edo castle in the Ichigaya neighbourhood is a stretch of water with quantities of greenery that is currently used only for recreational fishing, At present, the public is kept at a distance from the moat because of the four-lane road. This proposal calls for the removal of two lanes of Sotobori Avenue and the installation of a waterside promenade. The connection between the water and the urban area will also be enhanced if the two-lane road is designated for roadside services without encroaching on the through-traffic. The waterfront area created alongside the existing road can be used to develop commercial establishments, such as cafés and shops. One proposal for the arterial road that was served by Sotobori Avenue is to transfer it to a new road to be built above the Chuo and Sobu Lines of the railway that run south. Because this development will occur at the same level as the bridge that crosses the moat, there will be no need for an overwhelming overpass structure and it will not impose much influence on the existing roadside cherry-blossom trees along the embankment.

English translation by Norie Lynn Fukuda

NOTES

1 Data based on technical surveys conducted by Tokyo Gas Co. Ltd.
2 Data based on technical surveys conducted by Tokyo Gas Co. Ltd.

BIBLIOGRAPHY

Disaster Mitigation Community Development Planning Support System (Disaster Mitigation Community Development Planning Support System Committee, www.bousai-pss.jp, presented by Takaaki Kato).

Organization for Property Damage Insurance Fee Computation (2005) *Shigaichi Tokusei wo Kouryo shita Jishin-kasai no Ensho Hyouka Houhou no Kaihatsu*, ('The development of fire spread evaluation techniques caused by earthquakes in regard to urban characteristics').

Technical surveys conducted by Tokyo Gas Co. Ltd.

7 EXCAVATING THE LOST COMMONS

Creating green spaces and water corridors for eco-urban infrastructure

ISHIKAWA MIKIKO

INTRODUCTION

Various types of commons have existed all over the world in all periods of time. These commons were created in local areas from the interaction between ecological and cultural conditions and played an essential role in sustaining the communities surrounding them. As in other countries, the rapid urbanization experienced in Japan from the 1960s involved the wholesale destruction of the social and environmental structure of the commons located on the fringes of existing cities. This chapter discusses how one such lost commons has been regenerated in Kakamigahara City in Nagoya metropolitan area. When and why did people start to notice the importance of the commons? What kind of planning process was invented for creating social consensus?

The chapter traces ten years' experiences and clarifies the trial and error that was involved in the regeneration of the commons. The most important result of the project was that it produced a clear vision for the city for the next generation. The three corridors, the Forest Corridor, the River Corridor and the Urban Corridor, and seven cores were designated as constituting the new commons. Then, based on existing legal and financial constraints, priorities were determined. Precise action programmes were followed and progress was monitored. In the process of the actual planning and construction, many citizens were involved so they could quickly understand the result of their efforts and their contributions to the community. In my belief, the power of this regeneration project arose from the efforts of many citizens who did not speak out eloquently, but silently and willingly worked for their communities. This project received the Japanese Prime Minister's Award for Greening Cities in 2005.

LOCATION OF KAKAMIGAHARA

The city of Kakamigahara is located in the suburban fringe of metropolitan Nagoya (Figure 7.1). To the north of the city is the edge of the Mino Mountains, and to the south the River Kiso flows from east to west. Between them is the wide plateau where human settlement started more than 5,000 years BC.

Figure 7.1 Location of Kakamigahara City in metropolitan Nagoya.

From the seventeenth to the nineteenth century the main road from Edo to Kyoto, Nakasen-Do, passed through Kakamigahara and ships travelling up and down the Kiso River were the focus of active trade in this area. After modernization in the Meiji era (1868–1912), a military base was constructed in the midst of the Kakamigahara plateau, and related industries provided the origins of the technical engineering which is now the main industry of the city.

During the 1960s rapid urbanization took place in metropolitan Nagoya and the city also experienced a sudden population increase. The area at the foot of the mountain, which is called Satoyama woodlands, was converted into Dan-chi, a complex of condominiums and single family houses. Before this, Satoyama woodlands was a traditional Japanese commons. It had been owned by the community the adjacent village and since time immemorial these people had had the right to collect in Satoyama woodlands the wood and leaves that they used for fuel and fertilizer. They had maintained the forest by cutting and planting periodically since the Edo era in the seventeenth century. The essence of this system was a sustainable commons, which had now completely disappeared.

Satoyama woodlands was also a rich source of bio-diversity. The ecosystem that had been maintained by the interaction between humans and nature produced rich species diversity. It was such a well-ordered environment that no one realized the importance of the woodlands. In the 1960s energy revolution, the fuel consumption shifted from wood to gas and oil. The most important role of Satoyama woodlands as an energy resource had ended. Since to maintain Satoyama woodlands was hard work for the community, the simple solution appeared to be to sell them and give the revenue to the community. Hence, the huge woodlands disappeared in the 1960s, not only in Kakamigahara, but all over Japan.

In the 1980s, citizens started to notice that there was something wrong. The water level of the small rivers that flow in the community decreased dramatically when the forest in the upper river basin disappeared. More striking still was the pollution of the drinking water. Since the source of drinking water in Kakamigahara is underground, the loss of the commons and the heavy application of chemical fertilizers on the agricultural land that replaced them caused serious pollution. Finally, the importance of the Satoyama woodlands was understood.

CREATING THE WATER AND GREEN CORRIDOR PLAN

In 1999 the city developed a grand plan for next 20 years. This was a plan for a Water and Green Corridor based on the Green Preservation Law, one of the fundamental Japanese city and regional planning laws. The basic feature of this plan was that it should be easy for even children to understand, and it should clearly depict a vision of the city in the future. A method of soliciting citizen participation was introduced, as this was the first experience of participation for both citizens and the municipality.

Three corridors and seven cores were designated. The corridors were the Forest Corridor, the River Corridor and the Urban Corridor. The seven cores selected were to be intensive areas where strategic planning and design should be implemented (see Figure 7.2, colour insert 2). Also, precise policies were established to preserve natural resources, and for the arrangement of parks and open spaces and the greening of the city. If this plan ended here, it would be no different from hundreds of similar plans throughout Japan. The significant difference of this Water and Green Corridor Plan was that it was accompanied by an action programme and careful process management that has so far been highly successful. The commission established by the city for managing this process consisted of local citizens, academics and municipal officers. By regularly monitoring the achievements every year against the plan, it was easy to see which policies were difficult to implement, which were supported by the citizens, and whether the citizens could understand the actual result and process of the grand design.

STRATEGIC PLANNING FOR THE FOREST CORRIDOR

At first progress was very slow because no one knew where to start, or what should be done to achieve this long-term project. We decided to focus on the most urgent issue, the restoration of Satoyama woodlands. Figures 7.3–7.6 show the problems in Satoyama, which are still serious problems all over Japan, such as the excavation of sand and gravel and the illegal disposal of garbage and industrial waste, as well as the destruction of the forest by fire. One role of the university was to carry out an extensive

Figure 7.3 Sand and gravel excavation.

Figure 7.4 Illegal garbage dump.

survey of Satoyama, including its geography, vegetation and wild life, as well as the existing problems. In addition to surveying existing conditions, we surveyed the historical changes in land use of the Satoyama woodlands and found the cause of the rapid decrease in water flow and quality. We found that even though the destruction had a serious negative impact on Satoyama, there still existed rich bio-diversities in the remaining forest. The most important thing was to create a place where people can come and enjoy, and, through these experiences, can learn the real importance of Satoyama.

Figure 7.5 Forest damaged by fire.

Figure 7.6 Citizens participating in tree planting after a fire.

The natural heritage forest

We searched for an appropriate site for the restoration of Satoyama woodlands. We focused on the site of the dam for flood control that was under construction in 2002. The area was typical of the Satoyama woodlands, but a private developer had bought it to dispose of industrial waste. Since this was the watershed at the source of the Shinsakai River, the city bought it back to convert it into the flood-control dam. Figures 7.7 and 7.8 show the process of the restoration, and a small school for learning about Satoyama woodlands was opened. This school is now run by community volunteers and every week, diverse programmes are provided (Figures 7.9 and 7.10, see colour insert 2).

Figure 7.7 Changing the bottom of the reservoir from concrete to clay.

Figure 7.8 Existing reservoir in the Natural Heritage Forest (2007).

CREATING COMMONS IN THE MIDDLE OF THE CITY: THE URBAN CORRIDOR

We also understood the importance of creating a commons in the middle of the city. In the 1920s an agriculture school that had been the pride of Kakamigahara was established as an institute of the agricultural department of Gifu University. It was torn down and moved out in the 1980s. Left on the

Figure 7.9 Natural Heritage Forest (2007).

vacant land was a lot of huge trees that the community spontaneously started to count and protect with labels. Gradually, this community movement started to preserve the vacant land. However, in 1999 the city authorities planned to build a road through it, splitting the site into two parts, and to put the land on the market.

The Water and Green Corridor Plan in 2000 had as its basis a proposal to preserve and create a central park. Consequently, an action programme for a civic centre as one of the seven core areas proposed was developed in 2002. After careful research on traffic flows, the City Planning Committee finally decided not to go ahead with the road through the commons and construction on the central park was started. Since the city did not have enough funds for wholesale redevelopment, piecemeal works were started and still continue today. In 2003 the old welfare centre that was about to be torn down was renovated, and landscaping the surroundings started by preserving trees and creating a small water bio-tope (Figure 7.11). The revitalization of such old, unused public facilities enhanced the reputation of and support for this project. We then created a beautiful promenade utilizing the old ginkgo trees that remained on the edge of the Gifu Agricultural School site. We planted another row of ginkgo trees and created the promenade. In small Japanese cities grand promenades (Figure 7.12, colour insert 2) are seldom found. This became a very popular place for citizens to stroll in and Christmas illuminations have started to be lit there (Figure 7.13, colour insert 2).

Third, we preserved the important bamboo forest in the cliff-side, and started to create wide green lawns where children could play freely, in recognition of the importance of providing spaces for children to play in water and the natural environment (Figures 7.14, 7.15, colour insert 2 and 7.16). The park was named the Learning Forest by the community, since it was formerly the site of the agricultural school. The community collected acorns from this site and started to grow saplings to create woods for the next generation (Figure 7.17). While the Learning Forest was being constructed, distinctive

Figure 7.11 Old Welfare Centre, after conversion, with small bio-tope (2007).

Figure 7.14 Construction of lawns and the pond (2005).

changes gradually occurred. The Cyubu-Gakuin University decided to move to a site adjacent to this park, and opened in 2005. Since the city provides so-called 'garden parking' in the Learning Forest, it is easy for citizens to come into the old centre of city and enjoy strolling and shopping there. How to solve the decline of the old, central core of the city is a common issue in Japan. In the case of Kakamigahara, we met these challenges by creating commons in the midst of the city.

CREATING DIVERSE COMMONS

Creating the commons in the Forest Corridor and Urban Corridor were successful, so the city decided to expand this policy wherever renovation was

Figure 7.16 Volunteer citizens teaching children bamboo craft.

Figure 7.17 Raising young trees for woods for the next generation.

necessary. One of the good examples of this strategy was the renovation of the crematorium. In the Forest Corridor it was time to reconstruct the old crematorium built in 1970s. Close by was an old agricultural pond and the cemetery. The city decided to beautify the crematorium by incorporating the pond into the landscaping. Figure 7.18 shows the new crematorium opened in 2006, named the Meditation Forest. The regeneration of the adjacent forest has also started by planting trees, and this project has attracted international attention.

Figure 7.18 Meditation Forest (2006).

Figure 7.19 Cherry-blossom trees in Japan.

CONNECTING THE COMMONS USING THE RIVER CORRIDOR

Finally, the River Corridor connects these commons, forests and the grand Kiso River. This corridor is the core of the hydrological cycle and enhances biodiversity, also providing daily recreational space to the community. In Kakamigahara 1,000 cherry-blossom trees were donated by Kabuki actors in 1930, and since then this tradition has continued. The Shinsakai River is designated as being among the most beautiful 100 cherry-blossom scenes in Japan (Figure 7.19). The city is now expanding this policy from the foot of the mountains to the Kiso River with the help of volunteers. There are also many small canals for agricultural uses. An extensive project is ongoing to convert the canal into a more ecologically viable structure. Figure 7.20 (colour insert 2) shows the comprehensive plan for the next 20 years. It will need to be supported by many activities enlisting community participation.

CONCLUSION

In this chapter I have described our ten years' experience of how to excavate lost commons and create new commons for the twenty-first century. The following points contributed to our success. It was essential to have a grand plan for the future of the city. It was created in collaboration with the community and the municipality. A clear action programme was necessary for implementing the plan, step by step. In each phase, community participation was essential. Achieving prompt results was important, even though some were very small. This was the source of energy and mutual reliance between the citizens and the municipality. Finally, reuniting nature and the man-made environment based on the ecological context in each locality was the fundamental method for regenerating the city for future generations.

8 CONTINUITY AND DEPARTURE
A case study of Singapore's Nankin Street

HENG CHYE KIANG

INTRODUCTION

The city of Singapore has a short history of only 188 years, the first 140 of which were under the auspices of British colonialism. Much of the early layers of the city fabric have therefore been predominantly inscribed by a British planning sensibility. Such historically imported layers have likewise significantly impacted on the cityscape and public space in the central core of Singapore. This situation remained intact throughout the turmoil of the Japanese occupation in the mid-1940s. With the first flush of independence and the formation of the first non-colonial government in 1958, the legacy of the colonial footprint began to be modified by different agendas and priorities. A new democratic mandate reflecting political, economic and physical realities required the imposition of subsequent layers on the oldest parts of the city that built upon and modified the colonial layer, rather than obliterating it. These new layers may not only change the physical grain and texture of the city and its spatial extent, but in many cases, may allow whole new functional roles to be developed within the existing fabric.

The physical development of Singapore is also unique. Given the very small area of the city-state – some 699 km^2 – and the need to accommodate all the functions of a state within it, ranging from water sufficiency to defence, planning in Singapore has taken on a dimension of importance rarely seen in other states at a comparable stage of economic development. From the very early years of independence, planning of various kinds – economic and physical (such as industrial, transport and green and blue development), environmental and the like – were taken very seriously in this vulnerable island with few natural resources.

One of the main principles adopted throughout the last 40-odd years was that of sustainable development. Economic development was an imperative. Environmental sustainability was also of utmost importance, since the constraint of physical size meant that the few physical resources available must be put to good and sustainable use, while providing a decent quality of life to its residents. Since the mid-1980s the cultural dimension had become increasingly important. While the goal of cultural development was explicitly announced and even translated into legislation in the case of conservation as early as 20 years ago, its significance was further underlined when a focus on identity became one of the seven key proposals in the last concept plan, Concept Plan (2001).

Following the publication of Concept Plan 2001, Singapore's Urban Redevelopment Authority (URA) launched identity plans for 15 areas in July 2002 to retain places with a sense of history and identity. Whereas land-use planning had hitherto addressed the issue of identity only by identifying historic buildings for conservation, the new Identity Plan seeks to 'identify the unique qualities of these areas and suggest ways to retain and enhance these qualities and even activities' (URA 2002). More than mere conservation, it seeks to identify the charm of various places, the things that make them appealing to us (Singaporeans), and it uses planning to help retain and evolve these characteristics to give us a sense of belonging, rootedness and identity. In this way, areas of significance with their unique character will evolve and remain the key focus of community life as the island develops.

Although the Identity Plan was officially launched only in 2002, some earlier developmental efforts that started as early as in the 1970s, 1980s and 1990s encapsulated much of what was enunciated in the Identity Plan. The case of Nankin Street is a case in point. It is worth tracing its development over the last 40 years to show how, in response to various exigencies ranging from environmental upgrading to economic sustainability, the street has managed to become a spine of physical and cultural continuity amid vast physical changes. Its transformation also encapsulates the basic principles of sustainable urban development by increasing significantly its floor-to-area ratio (FAR) and combining a rich mix of uses. In the initial phase of transformation soon after independence, the continuity of social and economic relations in the area was maintained to some degree while the physical fabric made way for larger footprint developments. Although more recent urban-regeneration efforts still seek to increase the FAR by introducing coarser-grain high-rise buildings, a significant quantity of the old urban fabric was also retained for adaptive reuse, even as the original inhabitants were replaced by new users.

NANKIN STREET

One of the oldest streets in Singapore, Nankin Street's history dates back to the beginning of the nineteenth century when Singapore was claimed by Sir Stamford Raffles as a free port. Located in the centre of the city, the transformation of Nankin Street epitomizes much of the urban development of Singapore. The street's development can be periodized into four main phases. In the Jackson Plan of 1828, Nankin Street is clearly discernible as a main street stretching from Telok Ayer Street on the east to what is today's New Bridge Road on the west. Parallel to Chin Chew Street, Hokkien Street, Pickering Street and Cross Street, this was the place where an early community of Chinese settlers had gathered and where later immigrants from China congregated.

Chinese immigrants flocked to Singapore after its establishment as a free port. By 1836, 46 per cent of Singapore's total population, or 13,749 people, were Chinese, up from 25 per cent or 1,159 just 15 years earlier (URA). The area quickly took shape according to plan with shop-houses, temples and public buildings lining the streets. By 1836 most main streets in the Telok

Ayer area had been fully constructed. A map drawn by J.T. Thomson in 1846 shows that the main developed area was demarcated by Telok Ayer Street, the Singapore River, New Bridge Road and Pagoda Street. The first half of the nineteenth century had seen the planning and construction of Nankin Street, occupied mainly by Chinese.

The opening of the Suez Canal in 1869 was significant for the development of Singapore and its port activities. By the end of nineteenth century Singapore had become one of the largest port cities in the world. In 1897 the municipal commissioner's Jubilee address described it as

> a city of 200,000 inhabitants, one of the largest sea ports in the world, visited in 1896 by ships whose combined tonnage exceeded 8½ million tons, and is the collecting and distributing centre for all the vast trade of Southern Asian and the Eastern Archipelago.
>
> (Yeoh 2003: 35)

The booming economy attracted even more immigrants and the city's population mushroomed to more than 500,000 in 1931 (Yeoh 2003: 317). Although the settlement area was expanded to accommodate the increased population, the population density of the Telok Ayer area continued to increase, bringing about the deterioration of buildings and the living environment. A shop-house originally designed to accommodate one or two families had to pack in over 100 residents. Rooms were divided and subdivided into cubicles. The cubicles that were used by the bachelors who came to Singapore first, had to house entire families when their relatives came to join them. The living conditions in those packed shop-houses were noted by Professor W.J. Simpson, who was commissioned by the colonial government to investigate the reason for the high death rate in Singapore in 1906, as being 'destructive to health' (Yeoh 2003: 94; Simpson 1907: 11).

After the Second World War housing conditions deteriorated further due to the reduction in the housing stock as well as to the constant increase in the population. The inflow of immigrants and high birth rates in the 1950s further exacerbated this deterioration. The Rent Control Act of 1947 that had been introduced to protect tenants from unfair rents discouraged landlords from maintaining their properties and led to the further deterioration of the housing stock. An investigation conducted immediately after the Second World War by the Housing Committee showed that about 300,000 people were 'herded into about 1,000 acres in the heart of the city … and with numbers of large blocks of houses, often back to back, with densities of 1,000 or more to the acre' (Wong and Yap 2004: 12).

In Nankin Street the situation was even worse. Upper Nankin Street was once regarded as being a top priority for demolition by the Singapore Improvement Trust (Kaye 1960: 7). In the 1960s Barrington Kaye investigated the living conditions of Upper Nankin Street and wrote:

> The majority of these were divided internally on the first and second floors, into small cubicles; each may be housing a family of 7 or more people.… Many of them sleep on the floor, often under the bed. Their

possessions are in boxes, placed on shelves to leave the floor free for sleeping. Their food is kept in the tiny cupboards, which hang from the rafters. Their clothes hang on the walls, or from racks. Those who cannot even afford to rent a cubicle may live in a narrow bunk, often under the stairs.

(Kaye 1960: 7)

According to this report, over half (56 per cent) of the inhabitants of Upper Nankin Street lived in households occupying a single cubicle, 7 per cent were obliged to share a cubicle with another household and 4 per cent had no other accommodation than the whole, or part, of a bunk. There were also 103 persons who had only a 'moving' space, usually a camp-bed set up in the storeroom (Kaye 1960: 66) (Figure 8.1). Even the street was used as the extension of temporary housing. The street became the dining room, laundry room, living room and storage room.

During the second phase of its development from the mid-nineteenth century to the mid-twentieth century, Nankin Street had evolved from a newly built main street and the urban activity centre of a new free port into an overcrowded, insalubrious and dilapidated slum-like place, occupied mainly by the lower classes of society. The third phase of its development began in the 1950s, when the density of the central area peaked with about a third of the island's population, or about 360,000 people, living on 1 per cent of a total land area of less than 600 ha (Dale 1999: 231). Improving housing conditions became one of Singapore's most urgent tasks. The 1958 Master Plan thus made population decentralization a priority and later gave rise to the Urban Renewal Programme (Dale 1999: 117).

In 1966 the Land Acquisition Act was passed, empowering the government to acquire land on a compulsory basis and the work of the Urban Renewal Programme started. The whole central area was divided into 19 precincts, numbered by the priority attached to their redevelopment (Figure 8.1). Nankin Street was included in precincts S3 and S6, suggesting that this street would be redeveloped in two tranches.

In precinct S3, redevelopment was carried out immediately after land acquisition. The new development was called the Hong Lim Complex and Fook Hai building. It was a group of buildings covering three city blocks of shop-houses and two main streets (Upper Hokkien and Upper Nankin Street). In 1968 over 200 shopkeepers in shop-houses in the precinct were served notice by the URA resettlement department to quit their premises. Although they were offered compensation or alternative space elsewhere, most chose to undertake an urban-renewal project primarily for their own use, as they had been established in the area for many years. They formed Fook Hai Development Pte Ltd for this purpose. The URA allocated the 2,600 m^2 site bounded by Upper Hokkien Street, Upper Nankin Street and South Bridge Road in March 1970. The new development of 22,300 m^2 – Fook Hai Building, comprising a seven-storey podium of mainly shops and offices topped by a 12-storey tower of offices and apartments – was completed in 1976.

On the remaining three city blocks, a total of five 18 to 20-storey Housing and Development Board (HDB) flats atop four-storey interlinked podiums

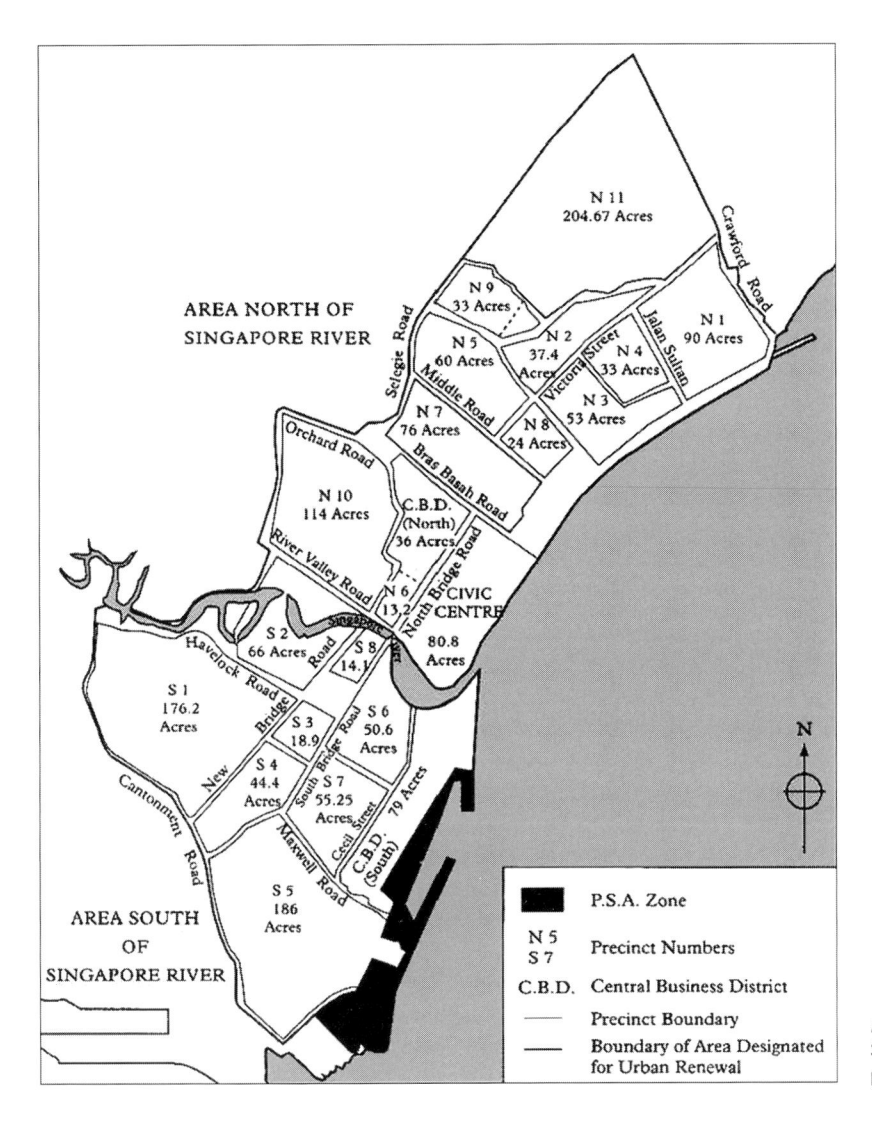

Figure 8.1 Central area of Singapore divided into 19 precincts.

were built between 1976 and 1979. Although all the shop-houses were cleared, Upper Nankin Street was kept as a pedestrian street in the complex (Figure 8.2a). Commercial facilities such as shops, restaurants and banks were located along the pedestrian street. While the urban tissue and grain have changed dramatically, the land use and nature of activities along the thoroughfare had persisted. Even though it is no longer called Upper Nankin Street, its physical form as a street with all kinds of activities remains as part of the Hong Lim Complex (see Figure 8.2b, colour insert 2). Today, the pedestrian street in the complex still flourishes with its vibrant activities and plays an important role in the public-space system of the central area.

Land acquisition for the rest of Nankin Street was conducted between 1974 and 1975. The shop-houses were not immediately cleared after all the residents had been relocated. This provided an opportunity for its subsequent redevelopment in the 1990s. The economic recession that took place in the

107

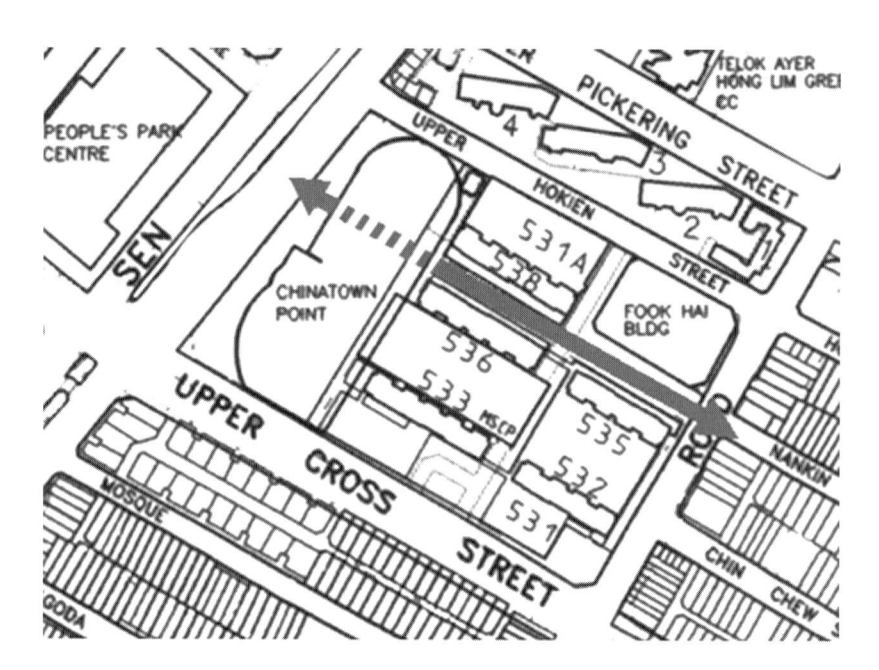

Figure 8.2a Map of the pedestrianized Upper Nankin Street.

first half of the 1980s seriously affected the direction of Singapore's urban redevelopment and, specifically, the redevelopment of Nankin Street. For the first time since 1967, Singapore experienced a GDP contraction of 1.4 per cent in 1984 (Dale 1999: 42). Economic woes were widespread. There was a glut of office and commercial space after years of rapid expansion. Tourism declined for the first time since 1965 and hotel occupancy rates fell from 86.1 per cent in 1980 to 55.1 per cent in 1986 (Singapore Tourist Promotion Board 1996). The construction of the HDB flats in the central area was halted and the development plan for HDB flats in the Nankin Street area was also cancelled.

Studies that were commissioned at the time pointed to conservation as a potential solution to the tourism sector, as well as to real-estate woes. The old shop-houses began to be seen less as insalubrious slums than as an opportunity to relieve the stress of economic recession. Several conservation schemes were thus announced. In 1984 the URA's conservation development proposal for the Tanjong Pagar area was approved by the government. In 1986 the Conservation Master Plan was announced. More than 100 ha of old Singapore were included in this plan. Several historic areas in the central area were gazetted as conservation areas, including Chinatown, the Singapore River, Kampong Glam, Little India, the Heritage Link and the Emerald Hill area (URA 1985–6: 2). Telok Ayer area was one of the four sub-areas in the Chinatown Conservation Area (Figure 8.3). Strict guidelines aimed to protect the historical building and the area atmosphere.

Although it was located in the Telok Ayer area, Nankin Street was not included in the Chinatown Conservation Area in the 1986 Conservation Master Plan. However, its proximity to the conservation projects provided another development possibility for Nankin Street after the original plan for development as HDB flats was aborted. When the conservation scheme for Chinatown Conservation Area was in progress in the early 1990s, the government started to look at the redevelopment possibility of the area bounded by Cross Street, South Bridge Road, Pickering Street, Church Street and Cecil Street. This area was named China Square and Nankin Street was one of its main streets.

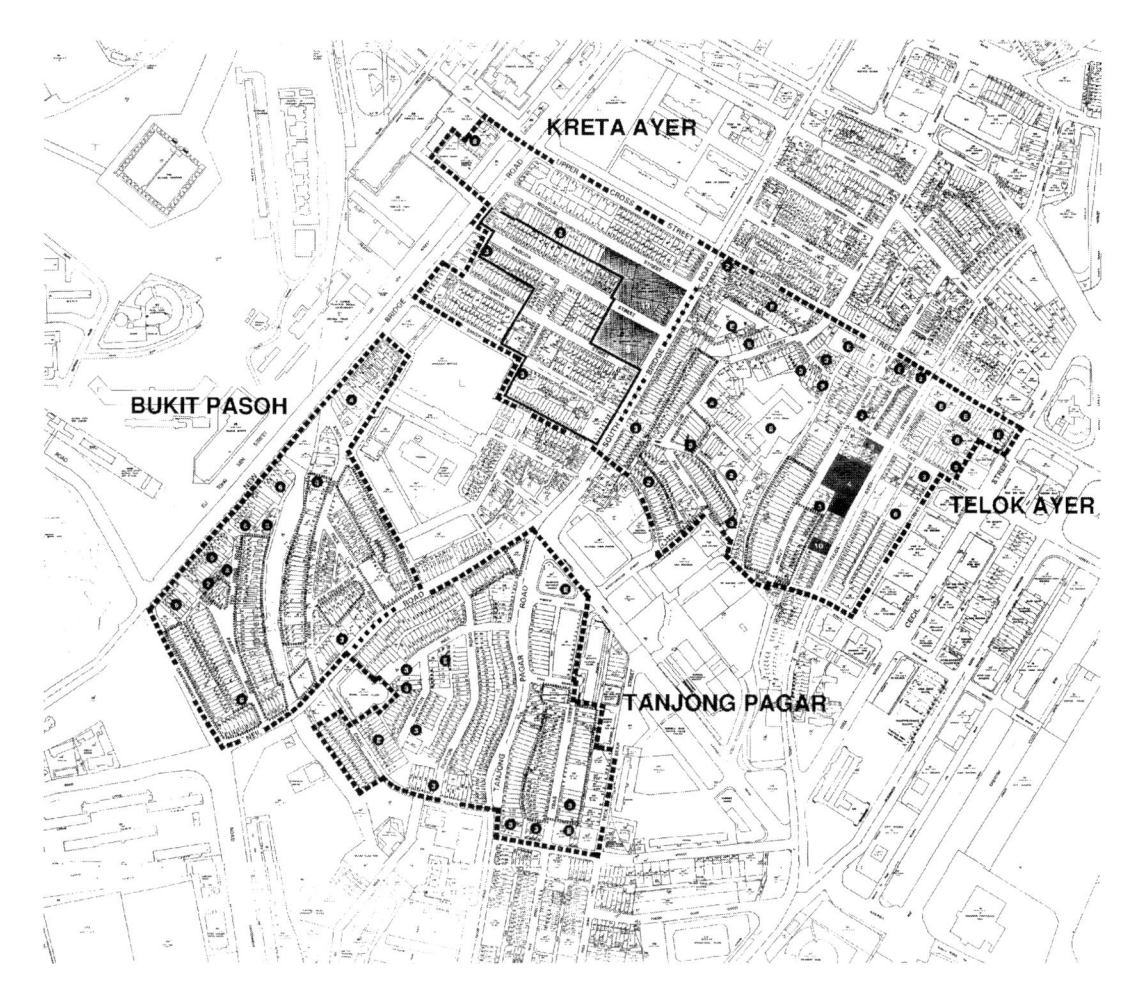

As the one of the last areas to be redeveloped after the land acquisition in the 1970s, the development of its environs provided unique new features for this area. Its location between the central business district of Singapore to the north and the Chinatown Conservation Area to the south provides China Square with a unique development opportunity to be a transition area between the CBD and the Chinatown Conservation Area. During the government planning studies, the historical and architectural value of the existing buildings were taken into account and weighed against the redevelopment potential of the land. The development approach that was finally decided was a combination of new development and selective conservation.

In the urban design scheme, about half of the existing buildings were retained and combined with vacant land parcels to accommodate large developments. Fortunately for Nankin Street, almost all the shop-houses lining it were selected for preservation. The whole area was divided into seven parcels,

Figure 8.3 Map of Telok Ayer in China Town Conservation Area (source: Urban Redevelopment Authority, Singapore, www.ura.gov.sg/conservation/telok.htm).

with Nankin Street lying between parcels F and G. While preservation of the shop-houses along the street would be completed by the developers of parcels F and G, the construction of Nankin Street as a pedestrian spine going through the whole area, was to be completed by the government.

In the whole urban design perspective, when fully developed, the China Square area was envisaged by the URA to become a vibrant activity hub forming a transitional zone between Chinatown, the Singapore River and the central business district. Nankin Street, together with Peking Street, would be important pedestrian linkages in this transitional zone. The planning and urban design scheme was unveiled by the URA in 1994. At the end of 1996 and the beginning of 1997, parcels F and G were sold via a land-sales programme. The successful tenders for both sites were submitted by the same developer. Hence, these two parcels were treated as a single development with Nankin Street as the public space going through it.

The project, which was completed in 2001, retained most of the shop-houses along Nankin Street and converted them to commercial use. Two 15-storey office towers were erected *behind* these shop-houses. A central plaza with a fountain was introduced on the south side of Nankin Street as a gathering place for the adjacent developments (Figure 8.4). The architectural design enhances a blend of old existing buildings and new development – the main concept in the urban design for the China Square area. The contrast between old and new was deliberately emphasized by the juxtaposition of modern office towers and traditional shop-houses. This intention was further underlined when a glass canopy was added in the place of shop-houses along the old Nankin Street to serve as a portal to the central plaza (Figure 8.5, colour insert 2).

Today Nankin Street has become a premium destination during lunchtime, evenings and weekends as most of its shop-houses were adapted to reuse as restaurants. It also plays the role as the pedestrian spine between the Chinatown Conservation Area and the CBD. With the help of an overhead bridge, the pedestrian link of Nankin Street extends beyond to the pedestrian-

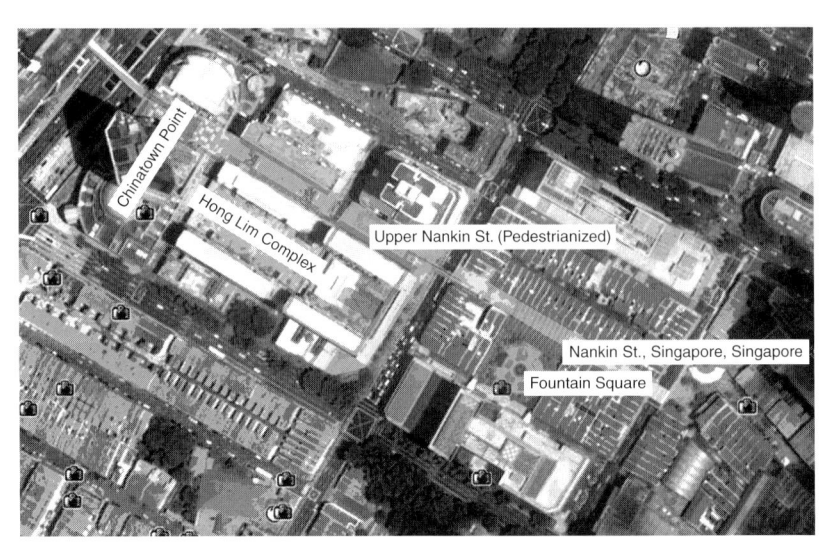

Figure 8.4 Satellite map of Nankin Street and the central plaza with a fountain in the centre (Fountain Square) (source: GoogleEarth 2008).

ized Upper Nankin Street in the Hong Lim Complex and finds its way through a commercial development (Chinatown Point) to New Bridge Road, just as it had been drawn and planned in the 1828 Jackson Plan.

CONCLUSION

Located in the heart of the central area, Nankin Street has witnessed and participated actively in the transformation of the city-state. Through the various phases, especially the last two, it has shown that physical and even social and cultural continuity is possible amid dramatic urban transformation. The development of Nankin Street throughout the last 40 years is a good case study of how the city has managed to retain physical and cultural continuity while intensifying land use, responding to economic exigencies and providing a better quality of life in general. Led by different agencies under changing developmental agendas over time, the transformation of the areas along Nankin Street engendered a radically new urban tissue along certain stretches while conserving and adaptively reusing the historical fabric at other sections. Both radical departure and physical, social and cultural continuity are present along the now pedestrianized street, itself part of a larger network of public space that has taken shape over the last two decades, during which new as well as old and familiar spatial and cultural practices of the city and its public spaces have evolved.

BIBLIOGRAPHY

Dale, O.J. (1999) *Urban Planning in Singapore: The Transformation of a City*, South-East Asian Social Science Monographs. Shah Alam, Selangor: Oxford University Press.

Kaye, B. (1960) *Upper Nankin Street, Singapore: A Sociological Study of Chinese Households Living in a Densely Populated Area*, Singapore: University of Malaya Press.

Simpson, W.J. (1907) *Report on the Sanitary Condition of Singapore*, London: Waterlow & Sons.

Singapore Tourist Promotion Board (1996) *Annual Report*, 1965, 1970/1–1995/6. Singapore: The Board.

Urban Redevelopment Authority (URA) (1988) *Historic Districts in the Central Area: A Manual for Chinatown Conservation Area*, Singapore: The Authority, p. 1.

Urban Redevelopment Authority (URA) (n.d.) Conservation areas and maps: Chinatown. Available online: www.ura.gov.sg/conservation/telok.htm (accessed 20 May 2008).

Urban Redevelopment Authority (URA) (1985–6) *Annual Report*. Singapore: The Authority.

Urban Redevelopment Authority (URA) (2001) *Concept Plan 2001*. Available online: www.ura.gov.sg/conceptplan2001 (accessed 20 May 2008).

Urban Redevelopment Authority (URA) (2002) *Identity Plan*. Available online: www.ura.gov.sg/pr/text/pr02–42.html (accessed 20 May 2008).

Wong, T.C. and Yap, A.L.-H. (2004) *Four Decades of Transformation: Land Use in Singapore, 1960–2000*, Singapore: Eastern University Press.

Yeoh, B.A. (2003) *Contesting Space in Colonial Singapore: Power Relations and the Urban Built Environment*, Singapore: Singapore University Press.

9 THE CULTURAL CHALLENGE FOR SUSTAINABLE CITIES
Coping with sprawl in Bangkok and Melbourne

SIDH SINTUSINGHA

This chapter investigates sprawl as a holistic phenomenon existing in a dialectical relationship with development elsewhere in the city. It explores the implications and repercussions through the broad framework of sustainability, urban sprawl and the responses in the inner parts of the city in laissez-faire Bangkok and planned Melbourne. In the process the chapter also raises further questions and implications for the cross-national comparisons that are increasingly necessitated by the current search for global consensus between developing and developed countries to mitigate climate change in an equitable manner (Figure 9.1).

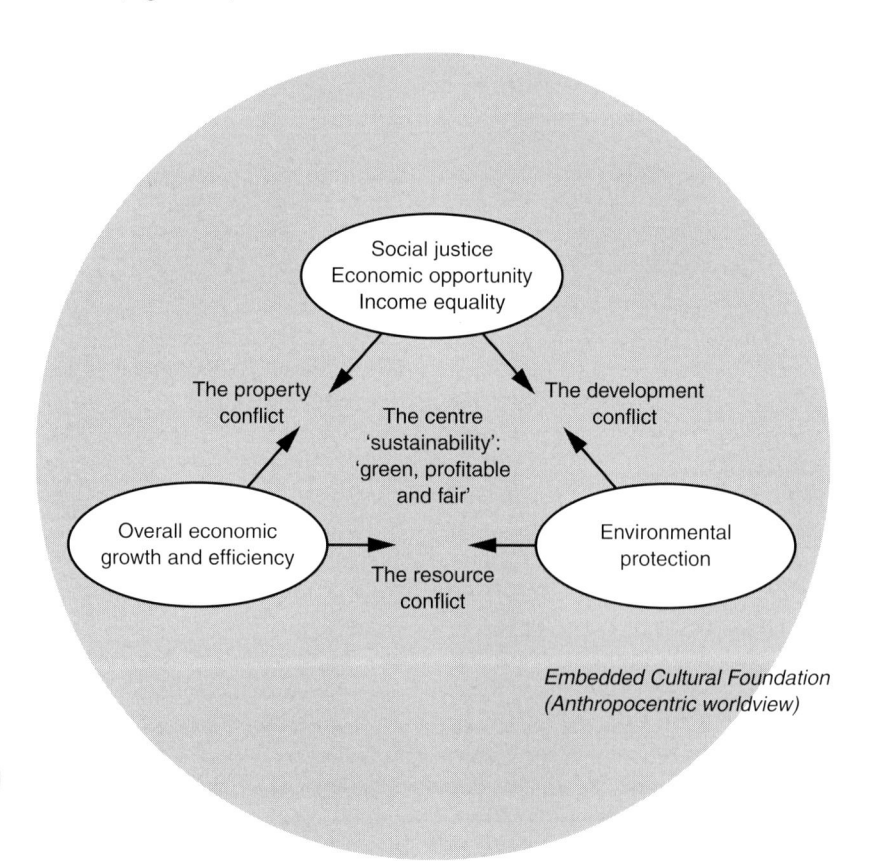

Figure 9.1 Planners'/designers' triangle modified from Campbell (1996, p. 298). Author addition: 'embedded cultural foundation'.

While sustainability has become a global catchphrase (Berke and Conroy 2000: 22; Campbell 1996: 301), the challenge still remains for it to establish true global depth and validity and avoid becoming an empty, or at least boundary-less, concept. To achieve that end in practice, sustainability needs to be flexible enough, in its meanings and forms, to address, balance and mediate between the unavoidable inconsistencies between anthropocentric and eco-centric world-views, comprehensive visions, incremental actions and scales of intervention through the diverse and complex spaces of cultural, socioeconomic differences (Sintusingha 2005) – a clearly onerous task.

However, if purposefully adopted, sustainability has the potential to be the broadest instrumental framework (Figure 9.1) for multiple negotiations across various scales of the economy, society, the environment and culture, as opposed to the current narrow global responses to climate change and the depletion of petroleum resources which mainly favour market (emissions trading) or technological fixes (cleaner fuels and increase in a new vehicle-fuel economy). In this context, sustainability functions as the, albeit voluntary, meeting place where synergies and compatibilities are identified and competing interests and conflicts are mitigated (Sintusingha 2005: 119). Due to these factors, the definitions and interpretations of sustainability cannot be rigid but must flexibly cater to both tangible and intangible differences across geographical spaces, whether at the scale of continents and countries with developed and developing economies or within cities between different socioeconomic groups. This is the focus of this chapter. Moreover, as a global meta-narrative, representing the more humane face of globalization, sustainability must be supple enough to engage responsively with multiple local cultures.

In the urban context, sustainability discourse has been biased towards the more established and developed Western city, rather than towards developing cities, which are growing at historically unprecedented rates (Sintusingha 2006: 151), posing challenges for future sustainability. Sustainability discourse is biased in favour of the middle to upper socioeconomic groups over the urban poor in the city. Another bias favours the central city and compact urban forms as solutions for urban sustainability, dismissing the suburbs that house a significant proportion of the city's population (and increasingly, the employment opportunities for them) as unsustainable and undesirable (Sintusingha 2007).

Yet in this globalizing world, while the homogenization of urban practices is one of the root causes of our current ills, at the same time globalization's communication technologies provide the means by which solutions that arise can be perpetuated, expanded and multiplied – it is hoped – into locally sensitive global best practices. However, the debate remains wide open between and within cities on who negotiates and decides where the balance is to be found between the local and global, as agency and political cultures are critical in this process. Resistance to aggressive, indiscriminating global market forces takes the form, more often than not, of uncoordinated, organic efforts and may not be straightforward; for example, from the design point of view, local traditional forms are now widely assimilated by commercial forces in the service of tourism (Sintusingha 2007).

To utilize sustainability as a framework, each city (and each city fragment) must be seen as an individual with its own peculiarities and eccentricities, a mixture of indigenous urban traditions and ecological characteristics that clash with an overlay of modern urban practices. A thorough analysis and understanding of the urban history of each city is critical in unravelling the adverse impacts of unsustainable practices. The specificities of each city have to be integral to the trajectory of the city's sustainable future development (Sintusingha 2006: 21).

In this context, urban sprawl poses some of the most critical challenges to urban sustainability and is a common global phenomenon tied in closely with economic and population growth. As the world enters the urban era,[1] the population increase in cities is likely to be absorbed into the sprawling suburbs, expanding the city's ecological footprint in the process. The challenge for urban planners and designers is thus to strike a balance between socioeconomic needs and conservation and resuscitation, not only of ecological systems, but also of rural ones. Moreover, urban sprawl is arguably one of the critical interstices of globalization, unsustainability and local culture, manifested in the form of pre-existing patterns that are erased and assimilated in varying ways by the expanding city. In order to move towards the objectives of urban sustainability, the on-ground realities of urban sprawl need to be proactively addressed and innovative sustainable retrofits targeted (Sintusingha 2005: 152). In other words, both the ecological and cultural remnants of place may provide possible cues towards contemporary sustainability.

Moreover, the challenge to urban sustainability in the broader narrative is the global consumerist world-class urban lifestyle closely aligned with economic growth and employment opportunities that continually draw people to the city. In this respect, the city has an inescapable bias towards the economic over social and environmental aspects, which forms a powerful narrative challenging the anti-materialistic principles and practices of sustainability (Sintusingha 2008). The very notion of world-class global cities, or possibly of the modern city itself, denotes the aggressive, homogenizing practices and images of globalization that threaten local cultural landscapes and practices, further contributing to cutting the social, environmental and cultural ties that characterize the narratives and practices of surviving pre-modern traditions. While I do not advocate using the past to solve contemporary problems, precedents and lessons can be adapted for present conditions, unless we view the modern city as an inevitability towards which all cities will evolve. In this context, the single most important challenge, as evident in the case of climate change, is much more environmental than sociocultural. On the ground, however, the notion of the modern city may, like the sustainable city, be idyllic, as the space of the city, allied with its sprawl, is wide enough to cater for both world-class shopping facilities and squatter settlements, as in the case of Bangkok.

This chapter explores the tensions, on scales from the everyday streets to broader urban patterns, between global and local urban forms and practices concurrent with emerging conflicts in formal responses as well as environmental and ecological issues in the city. These tensions and conflicts are physically manifested more clearly in Bangkok while they are relatively subtle in

Figure 6.1 Early vision of the 'Fibercity'.

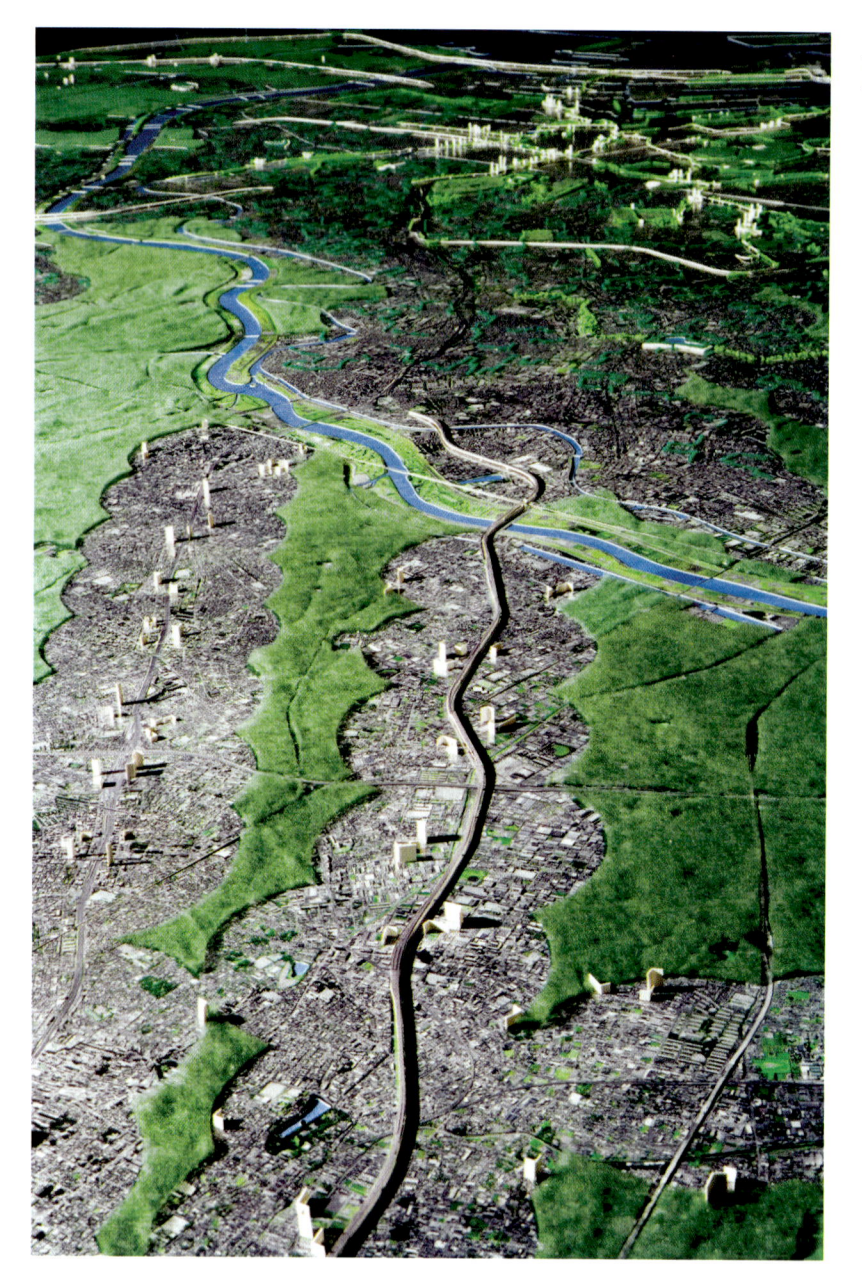

Figure 6.2 Modern photo of a Tokyo suburb reorganized using the Green Finger.

Figure 6.4 Improved housing area along Green Partition.

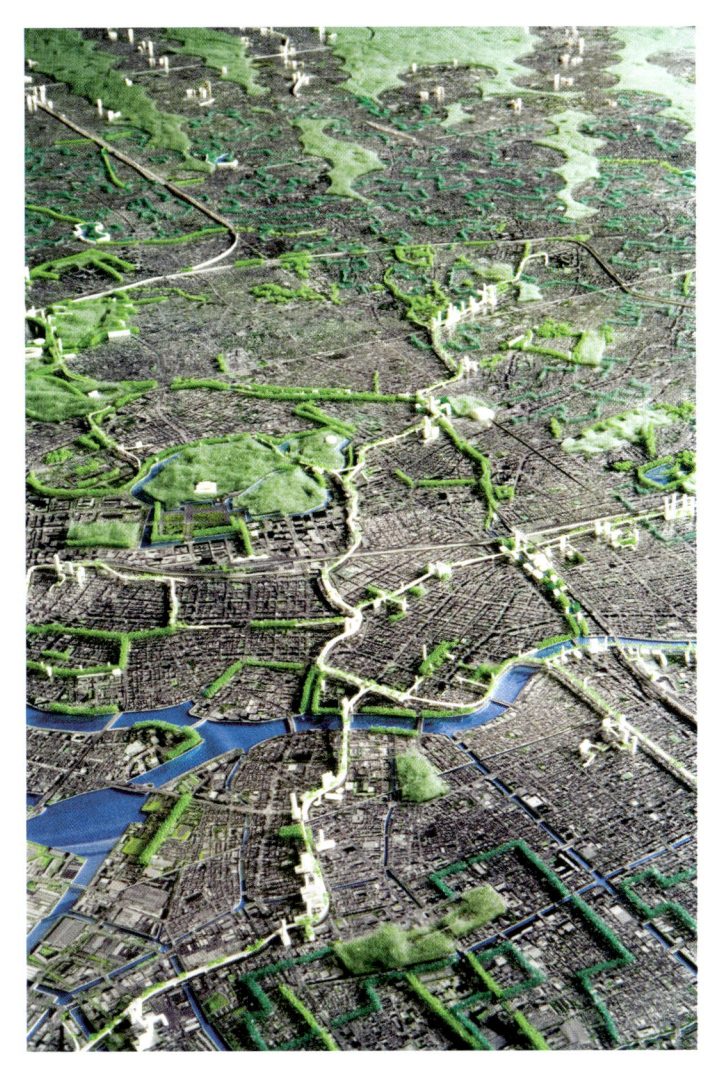

Figure 6.6 Model photo of the Green Web in central Tokyo.

Figure 6.7 Sumida Riverside.

Figure 6.8 Miyakezaka.

Figure 6.9 Three-dimensional urban section including conversion of Metropolitan Expressway.

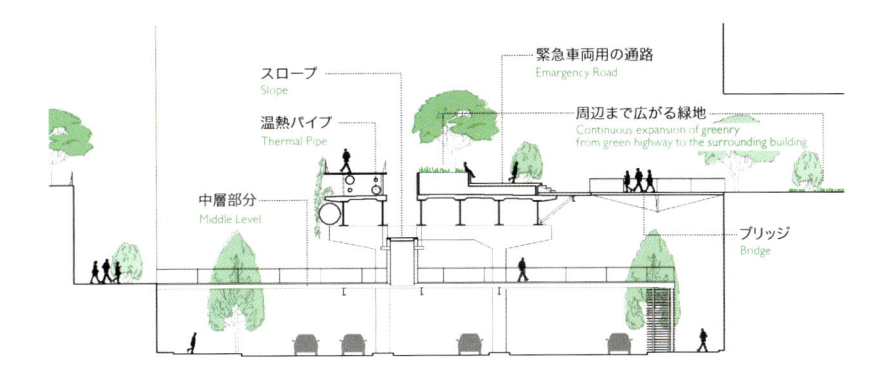

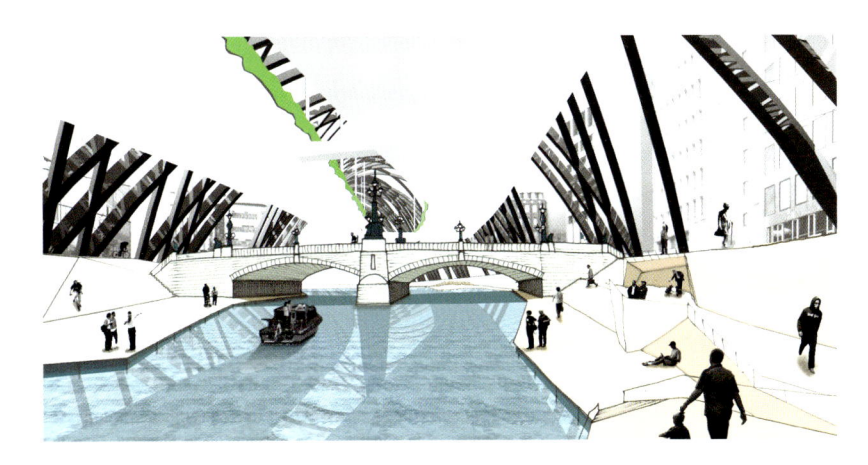

Figure 6.12 Intersection of two era-defining bridges.

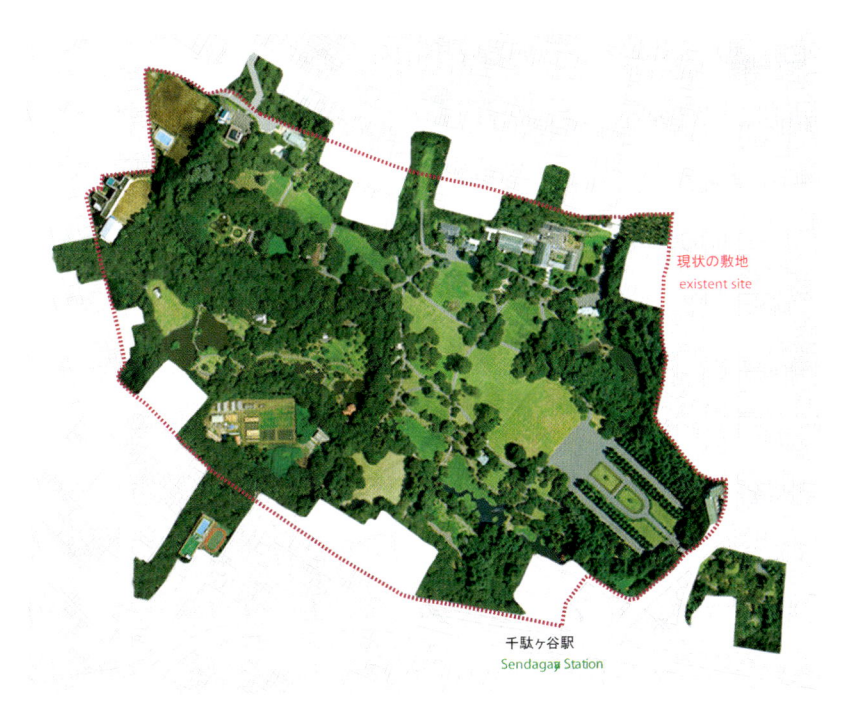

Figure 6.13 Making pleats on the boundary of the Shinjyuka Imperial Park.

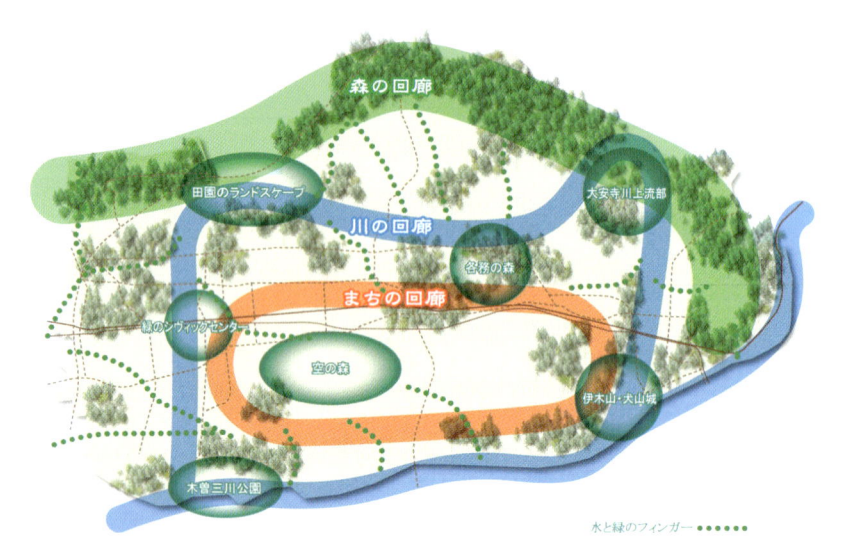

Figure 7.2 The concept of Water and Green Corridor plan.

Figure 7.10 Environmental Learning School.

Figure 7.12 Creating a promenade (2004).

Figure 7.13 Promenade at Christmas time (2006).

Figure 7.15 Citizen gallery and pond.

Figure 7.20 Water and Green Corridor plan in Kakamigahara for 2025.

Figure 8.2b Hong Lim Complex.

Figure 8.5 Glass canopy along the old Nankin Street.

Figure 9.2 Vertical separation: skytrain patrons cut off from vendor-lined streets.

Figure 9.3 Newly imposed bicycle lanes in Bangkok (left) and an upgraded counterpart in Melbourne (right).

Figure 9.4 World-class high-rise lifestyle 'terrorizes' Melbourne suburban idyll.

Figure 9.5 Uthaithani town is one of the very few places left where Bangkok's lost 'water-based' culture can be viewed.

Figure 11.6 Hill Council complex, Leh, India – bird's eye view.

Figure 11.7 Hill Council complex, Leh, India – section through council hall.

Figure 11.8 Hill Council complex, Leh, India – the main approach and front façade.

Figure 11.10 Hill Council complex, Leh, India – interior view of the building.

Figure 11.11 Hill Council complex, Leh, India – solar shells and light vaults.

Figure 11.12 Solar shells generate fascinating architecture.

Figure 11.13 Solar shells generate fascinating architecture.

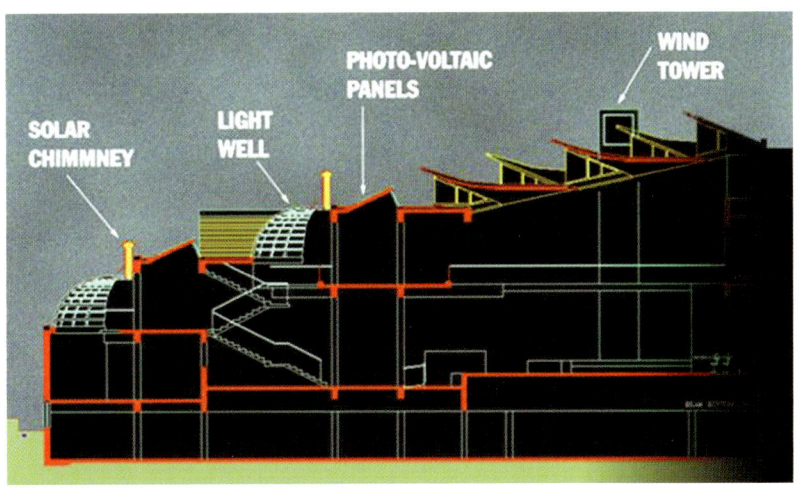

Figure 11.14 Hill Council complex, Leh, India – cross-section.

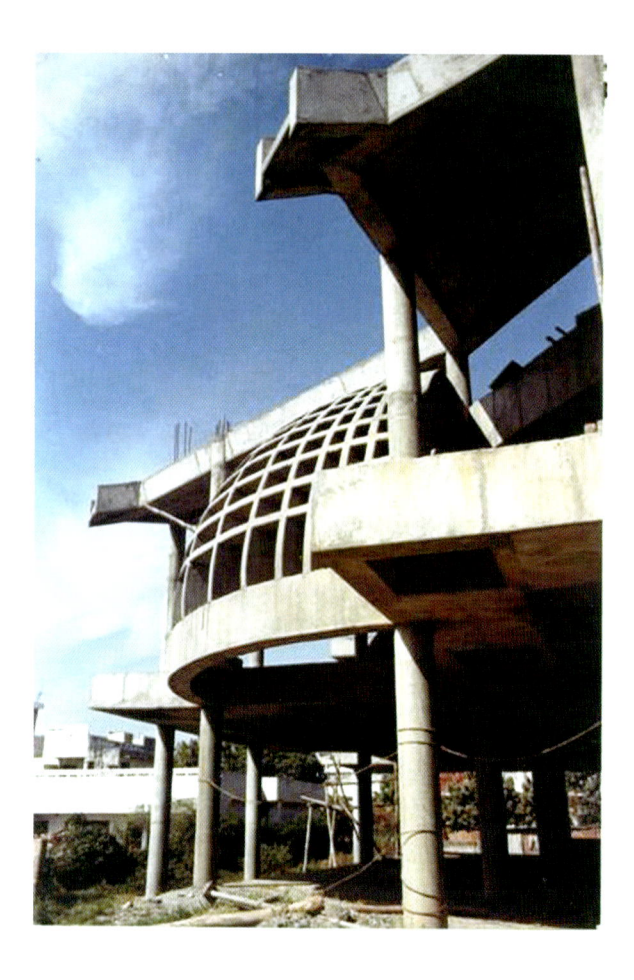

Figure 11.15 Hill Council complex, Leh, India – southern façade.

Figure 11.16 Hill Council complex, Leh, India – view of the entire complex.

Figure 11.17 Hill Council complex, Leh, India – sectional perspective.

Figure 11.18 Himurja Building, Shimla, India – building after retrofit.

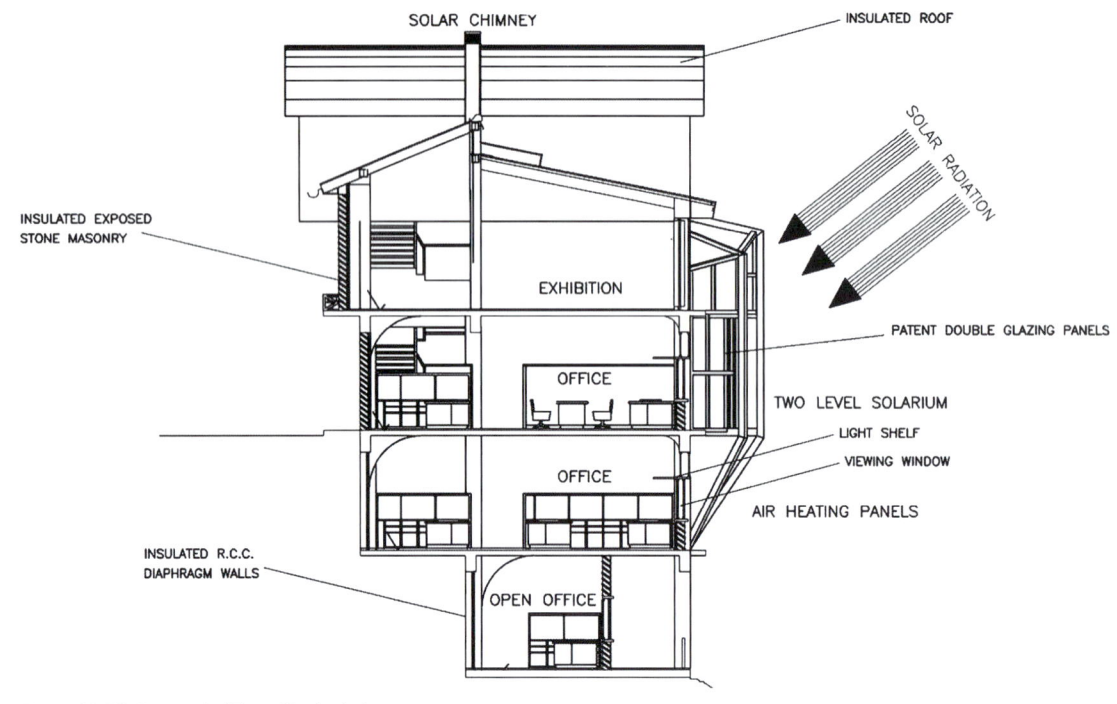

Figure 11.19 Himurja Building, Shimla, India – section.

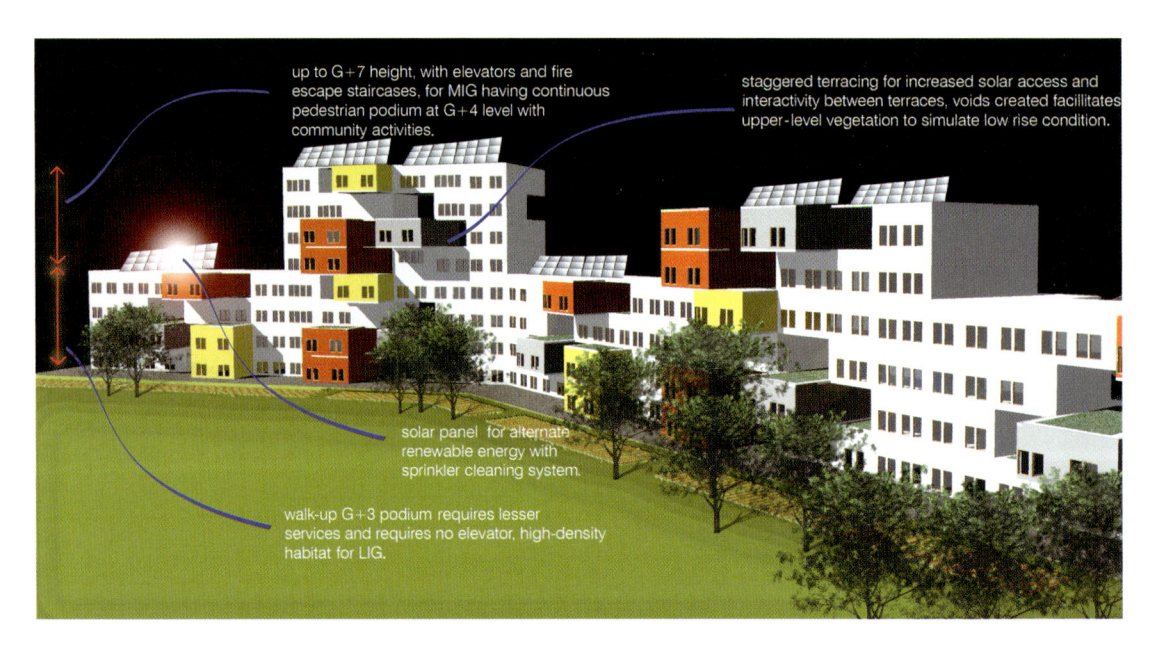

Figure 11.30 Gol Market, New Delhi, India – link habitat form.

Melbourne, reflecting and defining the status of their respective societies as developing and developed.

CONFLICTING URBANITY AND SCALES: BANGKOK'S STREET LIFE OR MEGA-STRUCTURES

> I live in Pratunam, and his comments mirror my own feelings. This area is slowly choking to a standstill due to more and more encroachment by vendors, the worst offenders being food purveyors. Very often the local post office is only accessible by walking crabwise past boiling oil and exposed braziers. Too many of the food vendors are simply filthy dirty and a public menace to health.... Irrespective of the corruption involved, the lack of supervision and control is totally inexcusable. The temporary banning merely confirms the authorities' awareness of the general squalor. My case rests.
>
> ... I thoroughly enjoy the street vendors and miss them on Mondays. The street vendors play a large part in making Bangkok the wonderful city that it is. Many of these people (they are people, you know) are working long, hard hours to support their families. If 'You You' is so compassionless and inconvenienced by their presence, then perhaps he/she should move from the lower Sukhumvit area back to where he/she came from.
>
> (*Bangkok Post* 2006)

Like other developing cities after the Second World War, through allied processes of industrialization and urbanization, Bangkok expanded spatially via roads and highways. Bangkok's growth was mostly laissez-faire in character, in a pattern of ribbon development clustering along the government-built main roads (*thanon*), incrementally branching off, via side-streets (*soi*) provided mostly by the private sector, into pre-existing villages, new housing developments and factories, erasing and imposing upon, yet mimicking (with speed being a critical differentiating factor) the canal (*khlong*) and side-canals (*khlong-soi*), the indigenous movement and transportation system (Sintusingha 2002). Bangkok's sprawl at the same time absorbed people from the overcrowded and unaffordable inner city and also displaced, transformed and modernized pre-existing densely settled rural villages in Bangkok's suburbs consistent with McGee's (1991) notion of '*desakota*', a Bahasa Indonesian term for 'village city'.

The economic boom of the 1980s to the mid-1990s accelerated this process and, inevitably, Bangkok's road infrastructure was found wanting. City administrators reacted by constructing raised expressways, new highways and ring roads, which only caused bottle-necks during the extended rush hours in the inner city's limited road network. They also planned to decentralize the government by moving various ministries to the outer parts of the city, all of which only exacerbated sprawl further beyond Bangkok's administrative boundaries. To encourage less car use in the inner city, the city administrators also planned rail mass-transit systems that were put into operation from the late 1990s with plans to eventually extend the network to the suburbs. This has significantly altered, and is still transforming, Bangkok's urbanity.

The construction of the sky-train and metro (initially servicing inner Bangkok) to mitigate the city's notorious traffic led to natural intensification through the mainly high-end residential towers and malls of various sizes along its routes above and below major roads, replacing the ubiquitous shop-houses and detached houses to emerge deeper into the *soi*s. This new phase or type of ribbon development resulted in the vertical spatial differentiation (Figure 9.2, colour insert 2) of socioeconomic classes. The spaces at street level are occupied by the indigenous vendor markets at transportation and activity nodes, and the commercial spaces on the raised stations are occupied by small, often franchised outlets, selling products similar to those in the streets below (in a case of derivative forms competing with authentic ones, both vending legitimate and pirated goods). This commercial utilization has not been replicated in the underground stations, which remain unoccupied despite spaces allocated for shops.[2]

The *thanon* and *soi* of Bangkok have always functioned as the city's civic spaces where differences converged and co-existed, continuing in the tradition of its fluid predecessor, the vast, intricate root-like system of *khlong* all draining into the Chao Phraya River, once the water-based[3] city's economic and cultural life blood. Pedestrian footpaths along the streets have a vibrant, almost chaotic character,[4] lined with mobile vendor markets that flexibly cater to a different clientele at different times of the day, whether they sell fresh produce, food, clothing and fashion accessories or tourist souvenirs.[5] They often form natural commercial synergies with shop-houses by the footpaths such as by offering complementing trades, leasing them space and supplying them with electricity and water. This contrasts with the newer malls and condominiums beside the streets which are regimented, gated, private fortresses. Many of their residents use private cars and mass-transit rail, viewing the street activities as a mess that needs to be regulated, reorganized or displaced. This is consistent with the city bureaucrats' views and policies, believing that the well-regulated practices of the original Western types of roads and footpaths should also apply. This is shown in Bangkok Metropolitan Administration's new policies that impose zones (such as those at Bobae Market east of Rattanakosin Island; Ribeiro 2005) to consolidate, regulate and limit vendor numbers and also in policies to encourage more bicycle usage to mitigate traffic and carbon emissions, with the bicycle lanes white-painted on the already busy footpaths, further competing with vendors for space (Figure 9.3, colour insert 2).

The middle to upper socioeconomic groups, now afforded the choice to leave their air-conditioned private vehicles at home (or at a park-and-ride facility) and the clogged streets for the speed and convenience of rapid mass transit, are living in spatial segregation from the streets. The deepening socioeconomic division[6] and entrenched attitudes (as expressed in the quotations at the beginning of this chapter) are further accentuated as the mass-transit system, particularly the Bangkok Mass Transit System (BTS) sky-train, have integrated commercial functions and activities, with franchised shops around the elevated stations and the additional connections via sky-walks to high-end malls realizing the commercial benefits of these connections.

The street-vendor markets, remnants of the fluid indigenous *khlong*, are viewed by many in the middle and upper socioeconomic groups as backwards, uncivilized and low class (as opposed to their own aspirational, world-class, global lifestyles), much like the squatter settlements in which many vendors live and where they are also subject to eviction, many to low-cost housing in the outer suburbs far from their employment in the city. Yet, their informal activities provides a livelihood and social mobility for these lower socioeconomic groups that form a significant proportion of Bangkokians and new migrants, which today include migrants from poorer neighbouring countries, to the city. In effect, both the informal, subsistent settlements and their economy[7] provide a critical socioeconomic role and mix (Sintusingha 2002: 139) that sustain Bangkok's urbanity.[8] A negotiated end to this tension is needed.

CONFLICTING SPATIAL CULTURES: URBAN VERSUS SUBURBAN LIVING IN MELBOURNE

> One of the world's leading architects is designing a Melbourne landmark for Docklands that will be Australia's greenest and most expensive office and housing complex.
>
> Iraqi-born, London-based Zaha Hadid will oversee design of a spectacular $1.5 billion scheme earmarked for Collins Street by Middle Eastern investment company Sama Dubai. Ms Hadid, 57, was the first woman to win architecture's most prestigious award, the Pritzker Prize, in 2004.
>
> *(The Age* 2007a)

In contrast to Bangkok, Melbourne's growth has been planned, expanding via an efficient grid. Additionally, rail public transportation, much of which was laid down before the mid twentieth century and consisting of complementary train and tram networks, has an extensive coverage and reach, particularly within 7km of the central business district, but with relatively low patronage (around 8.1 per cent of trips, according to the Australian Bureau of Statistics (ABS) 2006 census). Today the dominant mode of transportation in the city is by private car (around 65.6 per cent) and, with major road projects being planned and implemented, this is not likely to change. There is also an extensive cycle network that has been earmarked for significant expansion in a conscious move to more sustainability (*The Age* 2007b). Melbourne is also a city with a significantly lower density and, consistent with other Australian settlements, most housing stock consists of detached houses in individual plots (71.9 per cent of private dwellings, according to ABS 2006). In contrast, most Bangkokians live in shop-houses, row houses and medium-to-high-density apartments.[9]

With the rapid rise in real-estate prices from the mid-1990s to today housing has become increasingly unaffordable, particularly in the inner suburbs, so that new home buyers, either by choice (preferring the suburban lifestyle) or by default, are forced either into the outer suburbs and satellite cities and into less-desirable suburbs or to continue renting. House prices doubled across eight Australian capital cities, according to the ABS 'Established House Price Index, June 2004–June 2005' (Linacre 2007: 2).

Melbourne's median house price surged by 23.4 per cent in 2007 alone[10] (Schneiders *et al.* 2008: 8). The socio-spatial divide thus manifested between the city and the outer suburbs as well as between suburbs is determined by factors such as the distance from the central business district, the distance to public transportation and the distance from bodies of water and other amenities.

In response, the state government initiated the well-meaning Melbourne 2030 Plan in 2001, advocating urban growth boundaries defined by green wedges to limit sprawl, and encouraging intensification at designated activity centres, particularly at transportation nodes, reversing traditional suburban models. However, while the intensification of the inner city with residential medium-to-high-density developments such as that in the central business district and Docklands (Figure 9.4, colour insert 2) (and other precincts with views of water) may be a sustainable option to mitigate sprawl, this does not address issues of affordability and social diversity. Moreover, critics argue that while there are plans to increase the use of public transport, little has been done to expand the network or improve services to entice people from their cars.

On the other hand, new, cheaper, but significantly lower-quality housing stocks are being provided in high-density student apartments.[11] These apartments specifically cater to, and were driven by, the rapid growth of international students from Asia, who are used to a high-density urban lifestyle and who have been studying at Melbourne's universities from the late 1990s. This more or less coincides with the transformation of the central business district into a vibrant, internationalized precinct that increasingly resembles the 24-hour/day Asian metropolis and it is reflected in the changing demographic profile of the inner city. Of the 14,538 usual residents in postcode 3000 (the City of Melbourne), only 32.6 per cent were Australian citizens (ABS 2006 census).

Otherwise, most high-rise residential units, mainly located near the central business district or along the waterfronts, cater predominantly to upper socioeconomic groups favouring a particular urban lifestyle, unlike most Melbournians with their own house and backyard, representing a divergence in residential preference. Tensions between these two lifestyles occur at designated activity centres and transportation nodes where new medium-to-high-density residential units, in the name of the Melbourne 2030 Plan, are imposed upon pre-existing low-density suburbs, as Lewis warned:

> Finally, whatever policy may emerge from these considerations, it must be applied in a way that does not *terrorise, oppress* or *penalise* existing residents. Whatever may be the goals of our planning policies, they must not *expropriate people's rights* or *devalue their lifestyles.*
>
> (Lewis 1999: 260; emphasis added)

Here, many residential groups are opposed to medium- and high-density developments in their previously low-density suburbs, arguing instead for the heritage conservation of the character of their neighbourhood. The implementation of *Melbourne 2030* has thus been problematic, with major issues and cultural conflicts challenging sustainability revolving around affordability and density/character.

Melbourne 2030 is being used as a cover for all developments that push the envelope. There's anxiety in the inner suburbs because higher-density living is being proposed everywhere, not just near activity centres. Almost every suburb now has its own protest group.

(Newton-Brown 2007)

In contrast to the manifestation of Bangkok's social-class mix in a physical form, these conflicts are not obvious at the everyday scale. In Melbourne at the level of streets, squares and public spaces, activities are clearly demarcated according to their function and transgressing these uses is relatively rare.[12] With a much lower density, competition for space seems less of an issue, which is not so in practice in everyday urban culture or at the broader scale of abstract planning.

DISCUSSION: DEVELOPING OR DEVELOPED, ORGANIC OR PLANNED

Attempting comparisons between cities, especially cross-national comparisons between cities in developing and developed countries, is problematic. To start with, the socioeconomic gap is much more significant in developing Bangkok than in Melbourne, which could be considered a predominantly middle-class city. Although official numbers show that the differences in their respective population may not be large (around four million for Melbourne and over five million for Bangkok), the unofficial population of Bangkok, including those in conurbations in surrounding provinces and rural migrants who are still registered as residents in their provinces of origin, is at least twice that of Melbourne's. However, because of our globally shared environmental problems, a global framework is required to allocate responsibility equitably to reduce carbon emissions. Thus we need to be able to apply comprehensive and comparable urban-sustainability principles.

Melbourne has an all but regulated, low-density culture that seems to be a major barrier to mitigating sprawl. In contrast, Bangkok has no inhibition in adding condominiums and malls even if this places heavy burdens upon the limited infrastructure and transportation network, particularly the roads. This often has the effect of also attracting economic migrants who are absorbed into informal and semi-formal squatter settlements. However, as with Melbourne, the formal inner-city developments mainly cater for higher socioeconomic groups that are attracted to pre-existing amenities. As these developments are given priority in Bangkok, roads are often widened in response to the increased traffic, threatening and displacing street life in the process. These differing cultural foundations also manifest themselves in the practices of built form.

Although Bangkok has come a long way in terms of creating building codes and planning to control development (its first official master plan was enacted in 1992), it still has little or no urban design. Regulations govern what can be built on each individual plot but little attempt is made to coordinate design and scale or provide open spaces and views on the collective level. The result is an occasional architectural feature (such as the traditional forms of temples that have been concealed for a long time behind shop-houses), but beautiful streetscapes and districts are rare.[13] Moreover, there is little consideration for the human aesthetic experience at the urban scale; not just for people living in buildings adjacent to the high-rise apartments and low-rise houses, but also for the person on the street. The best views are artists' impressions or taken from the high vantage points that constitute views reserved for wealthy citizens. This may explain why many Western design commentators say that the urban architecture in Bangkok is poor,[14] some even remarking that Bangkok has no great modern architecture.[15] This can be partially countered, such as by the statement about vendors at the beginning of this chapter, that Bangkok compensates abundantly for aesthetics with creative mixed uses and the vibrancy of its street life (or its messiness, depending on how one views it). These unflattering views may also be due to culture and language barriers and the eccentric Thai brand of modernism that is sometimes compared to the linguistic style of a foreigner speaking broken English.[16]

This contrasts with Melbourne's urban tradition where urban architecture, through building infills and additions, high-rise residential blocks and major public buildings, contributes to a more or less harmonious collective public experience of the city for pedestrians, cyclists, motorists or those using public transport. However, ironically, it is this very culture of urban architecture and the associated urban/neighbourhood character that it has created that hinder the higher-density development essential to contain sprawl. Moreover, the high-end demand for aesthetics also prices out affordable higher-density developments in the inner city.

Broadly speaking, Melbourne and Bangkok are grappling with parallel issues, which are manifested in different ways. The issue of urban aesthetics, allied with the narratives of a world-class 'liveable city', dominate in Melbourne, a developed city, while it is more muted in developing Bangkok, with conflicting socioeconomic worldviews that are physically and visually manifested. For the globalized urban middle classes of both cities, the city's image – tied in with their aesthetics and lifestyle rights, and even more recently, their image of sustainability – seems to take priority over its actual sustainability.[17] The image of being a world-class and developed city takes priority to the content of development. Does the ultra-modern Suvarnabhumi Airport, for instance, make Thailand itself any more modern?

ECO-URBANITY?

The discussion of urban forms and patterns inescapably revolves around the socioeconomic component of sustainability, because it involves and affects the city and its citizens. What is often absent in discussions about the city is the ecological component of sustainability, which is often assumed to be a totally separate issue (Nicholson-Lord 1987; Spirn 1984), reinforced by the superficial separation of culture and nature. In that way, the concept of eco-urbanity proposed in this book to define, understand and plan for contemporary lived urban spaces, is an oxymoron, much like the concept of sustainable development. Yet, both are critical concepts and frameworks in reconciling what are perceived as opposing and separate realms to the detriment of genuine sustainability. On the other hand, as Maclaren (1996: 301) argued, we are unavoidably anthropocentric and our cities are still striving for anthropic urbanity (as shown in the examples used in this chapter), in the context of the dominant capitalist economic urbanity. Global environmental problems, in particular the impacts of climate change, may prod society to search for compromises and solutions. However, the solutions proposed have mainly been either technological or market-oriented. Until eco-urbanity is an integrated lifestyle and culture, short-term fixes may only yield and compound environmental problems for future generations.

The increasingly pertinent issue of climate change and the broad negotiations towards an acceptable global consensus to address it also pose further challenges on translating this knowledge in terms of city forms and cultures. While it may be necessary to impose blanket planning, design and technological solutions based on scientific research, if these processes are compounded and possibly hijacked by market forces, will this lead to another homogenization of city forms? Any cultural loss in the rush to modernize will be irreparable, even though it is also often hard for those within a culture to recognize and value their existing cultural assets (Figure 9.5, colour insert 2). In this respect Singapore makes interesting comparison with Bangkok:

> As early as the 1940s, when Singapore was still a British colony, efforts began to eradicate the 'scourge' of street food vendors, which was quite a typical municipal response at the time (McGee 1971) ... the only street restaurants that exist today in the city are in the gentrified areas that attempt to recreate the ambience of 'Old Singapore.' They are expensive and poor imitations of the coolie stalls of the past and mostly serve middle-class consumers and tourists.
>
> (Yasmeen 2001: 15)

A similar argument can be made by advocates of Melbourne's low-density suburban lifestyle, reflecting the complexities inherent in the difficult move to urban sustainability and eco-urbanity. In the midst of these cultural conflicts, it is hard to see the broad picture: that cultural loss, for good or for bad, is also marked by cultural gain (Figure 9.6).

Figure 9.6 Intersection within a neighbourhood block in Tokyo at night where vending machines assumed the role of vendors.

NOTES

1 When, for the first time in history, most of the world's population lived in cities.
2 There may be several reasons for this. A cultural fear of the underground was exacerbated by a train accident in its first month of operation, and there is no planned integration of the underground with commercial buildings such as malls and department stores. The network is still relatively limited and hence lacks a critical mass of users. However, on the last two points a similar argument could be made for the Bangkok Mass Transit System (BTS) sky-train.
3 I borrow Jumsai's (1997) term to differentiate Bangkok's indigenous water-based cultures from imported Western and Chinese land-based cultures.
4 As with most Bangkok urban eccentricities, the footpaths are under the jurisdiction of a bureaucratic organization separate from the roads – the *thedsakij* ('footpath police').
5 See Kasama Polakit's (2004) PhD thesis for an excellent account of Bangkok street life.
6 This social division between the urban middle class and the urban and rural poor manifested itself in the recent referendum for the new Constitution and the December 2007 general election.
7 Evers and Korff (2000: 33) prefer to use the description 'urban subsistence production', as they argue that 'the social economy of the poor urban masses is neither a market of subsistence economy, nor does it neatly fall into the formal or informal sector'.
8 One of the reasons why the 1997 financial crisis did not result in more extreme social crisis is that the urban poor had sources of food to fall back upon, whether through urban farming or by returning to their rural villages.
9 Officially, the density of Melbourne's inner suburbs is significantly less than that of Bangkok's. For example, Klongtoey has ~9,639 people/km^2 and Prakanong has ~7,075 people/km^2, while Stonnington has 3,508 people/km^2 and Monash has 1,978 people/km^2 (Bureau of Registration Administration, Ministry of Interior, Bangkok 2005 and Australian Bureau of Statistics 2006). Bangkok's unofficial population is much higher than is represented in these figures.
10 House price appreciation differs significantly from city to city and within cities. Australia's real-estate growth is parallel with growth in America and Europe.
11 This is being investigated through an Australian Research Council-funded project entitled 'Transnational and temporary: place-making, students and community in Central Melbourne', by a Melbourne University team (K. Shaw, R. Fincher, P. Carter, and P. Tombesi).
12 It is often enforced by the residents. I have been reprimanded on several occasions when riding the bike on the relatively empty footpath instead of on the designated cycle lane on the road.
13 With the exception of the old urban core where a conservation master plan influenced by the City Beautiful movement is being implemented. Again, the urban poor are displaced as part of the beautification process (Prakitnontrakan, 2007).
14 Bill Bensley, landscape architect based in Bangkok (Junker 2007: 72).
15 Inferred from the observation in the *Bangkok City Guide* (Cook 2006: 73) that Thailand does not have a 'great modern architect'.
16 Recent, more studied investigations, such as Ching *et al.* (2007) are more accommodating, stating that 'Thailand gives us a glimpse of what modern "eastern" architecture looks like that was neither colonized by the Europeans nor closed off in the name of tradition' (p. 583) adding that

whereas the architectural traditions began to dry out or became fossilized with the introduction of European-style buildings in India, for example, these traditions developed with a sense of freedom in Thailand, unfettered by the offended eyes of colonial overlords.

(p. 626)

17 See Nasongkhla's (2008) PhD dissertation on Mae Hong Son for a vivid account of this on a small tourist town.

BIBLIOGRAPHY

Australian Bureau of Statistics (ABS) (2006) *Census*, Canberra: ABS.

Bangkok Post (2006) Available online: www.bangkokpost.net/News/14Jun 2006_news99.php (accessed 14 June 2006).

Berke, P. and Conroy, M.M. (2000) 'Are we planning for sustainable development?', *Journal of the American Planning Association* 66, 1: 21–33.

Bureau of Registration Administration (2005) 'Area and population in Bangkok metropolis by districts in 2005'. Available online: www.bma.go.th/bmaeng/population/population.htm (accessed 17 May 2008), Bangkok: Ministry of Interior.

Cambell, S. (1996) 'Green cities, growing cities, just cities? Urban planning and the contradictions of sustainable development', *Journal of the American Planning Association* 62, 3: 296–312.

Ching, F., Jarzombek, M.M. and Prakash, V. (2007) *A Global History of Architecture*, Hoboken, NJ: J. Wiley & Sons.

Cook, R. (ed.) (2006) *Wallpaper City Guides: Bangkok*, London: Phaidon.

Evers, H. and Korff, R. (2000) *Southeast Asian Urbanism: The Meaning and Power of Social Space*, Munster, NY; Singapore, Lit Verlage: St Martin's Press, Institute of Southeast Asian Studies.

Jumsai, S. (1997) *Naga: Cultural Origins in Siam and the West Pacific*, Bangkok: Chalermnit Press and DD Books.

Junker, U. (2007) 'Local knowledge: Bangkok', *The Australian Financial Review Magazine*: 71–2.

Lewis, M.M.B. (1999) *Suburban Backlash: The Battle for the World's Most Liveable City*, Melbourne: Bloomings Books.

Linacre, S. (2007) 'Wealth in homes of owner-occupier households', *Australian Social Trends 2007*, Canberra: Commonwealth of Australia, p. 10.

McGee, T.G. (1991) 'The emergence of Desakota regions in Asia: expanding a hypothesis', in Ginsburg, N., Koppel, B. and McGee, T.G. (eds) *The Extended Metropolis: Settlement Transition in Asia*, Honolulu: University of Hawaii Press, pp. 3–35.

Maclaren, V.W. (1996) 'Urban sustainability reporting', *Journal of the American Planning Association* 62, 2: 184–203.

Nasongkhla, S. (2008) 'Aesthetics and change in the Tai cultural landscapes of Mae Hong Son, Thailand', Unpublished PhD thesis, Faculty of Architecture, Building and Planning, Melbourne, University of Melbourne.

Newton-Brown, C. (2007) in Pallisco, M., 'Shaping the future Melbourne', *The Age*, 28 July, Melbourne, pp. 4–5.

Nicholson-Lord, D. (1987) *The Greening of the Cities*, London and New York: Routledge & Kegan Paul.

Polakit, K. (2004) 'Bangkok street life: transformation of place in three communities', Unpublished PhD thesis, Faculty of Architecture, Building and Planning, University of Melbourne.

Prakitnontrakan, C. (2007) 'The area surrounding Mahakarn Fortress: from preservation issues to key political problems', *Association of Siamese Architects Journal* 2–3: 81–93.

Ribeiro, G. (2005) 'Research into urban development and cognitive capital in Thailand', *Journal of Transdisciplinary Environmental Studies* 4, 1: 1–5.

Schneiders, B., Rood, D. and Austin, P. (2008) 'Brumby warns on mortgages: House price boom "unsustainable"', *The Age*, 26 January, Melbourne, p. 8.

Sintusingha, S. (2002) '*Muang, khlong,* and *thammachat* (city, canals, and nature): ingredients for "sustainable" Bangkok', in *Modernity, Tradition, Culture, Water: Proceedings of an International Symposium,* October 2002, Bangkok. Bangkok: Kasetsart University Press.

—— (2005) 'Sustaining sustainability and its inconsistencies', *Ranaeng: Journal of the Faculty of Architecture, Kasetsart University* 4: 118–35.

—— (2006) 'Sustainability and urban sprawl: alternative scenarios for a Bangkok superblock', *Urban Design International* 11, 3/4: 151–72.

—— (2007) 'Erasure, layering, transformation, absorption: convergences of the local and global in the Thai cultural landscapes', in Bull, C., Boontharm, D., Parin, C., Radović, D. and Tapie, G. (eds) *Cross-cultural Urban Design. Global or Local Practice?* London: Routledge: 29–33.

—— (2008) 'Sustainable world-class cities and global sprawl in Southeast Asian metropolitans', in Jenks, M. Kozak, D. and Takkanon, P. (eds) *World Cities and Urban Form: Fragmented, Polycentric, Sustainable?* London: Routledge.

Spirn, A.W. (1984). *The Granite Garden: Urban Nature and Human Design,* New York: Basic Books.

The Age (2007a) Front page. 2 August.

The Age (2007b) Parks Victoria graphics. 4 August.

United Nations (2006) *World Urbanization Prospects: The 2005 Revision,* New York: United Nations Publishing.

Yasmeen, G. (2006) *Bangkok's Foodscape: Public Eating, Gender Relations, and Urban Change,* Bangkok: White Lotus Co.

10 GEOMETRIES OF LIFE AND FORMLESSNESS
The theoretical legacies of historical Beijing

JIANFEI ZHU

After three decades of rapid development, China today faces urgent problems that have accumulated in the recent past. These include an expanding financial and social disparity between people, the loss of public goods such as a clean environment and social welfare (housing, health and education) and the destruction of historical urban fabric. In the last instance, this includes not only the physical demolition of courtyard houses and laneways, but also the erosion of the overall quality of urban public space that was previously humane, habitable, accessible and green. Although current government policies aim to correct the one-sided development with many regulations issued to protect by force the environment and some aspects of the social fabric and public service, the critical, informed understanding of urban design in the Chinese context is still missing. For example, there has been a call to make Beijing humane (*yiren weiben*) and habitable (*yiju*), yet a body of critical knowledge for urban design in the Chinese context is wanting. How can we reconstitute for the general public a habitable city that has humane and ecological qualities? How can we imagine a new urban public space displaying these qualities that is modern and yet eminently suitable for China? How should modern Beijing evolve, taking along with it useful legacies from its ancient past and cultural traditions? Can traditional Beijing with a 500-year history teach us anything critical and constructive today? Can we situate a discourse on Beijing in an international context so that the project contributes to global dialogue and debate? To offer a preliminary answer to these questions, I first employ European theorists' readings of Chinese and Asian urbanities as mirror reflections on the Chinese situation, before turning to traditional Beijing directly, on which I will make four observations.

EUROPEAN REFLECTIONS

Henri Lefebvre has differentiated between 'conceived' and 'lived' space: one a dominant space of rational knowledge and political ideology, and the other a dominated space of our direct life experiences, in which our desires are expressed through images and symbols (Lefebvre 1991: 38–9). Although Lefebvre tended to see this as a universal difference, he also acquired a relativist viewpoint. While he believed that there was a distinction between these spheres, and that in Europe, especially since the Renaissance, the first dominated the second, he wondered if such a distinction existed in Asia and China, where a different writing system had been adopted. According to him, in post-Renaissance Europe the space of the conceived, represented by the use of perspective (the gaze and the logic of visualization), the high façade, the front of buildings as aesthetic objects, and the long avenues leading to them, dominated and were clearly different from the textured medieval and religious space of the city's life-world (the ego, bed, bedroom, dwelling, house, square, church, graveyard, and the symbolic figures of devils and angels, heaven and hell) (Lefebvre 1991: 41–2, 47). Lefebvre then asked if such a distinction existed in Asia and China. Chinese characters, being different from alphabetic writing, acted as a crucial sign in Lefebvre's thinking, pointing to another possible world without such a distinction (Figure 10.1). 'It is indeed quite possible', he argues,

> that the Chinese characters combine two functions in an extricable way, that on the one hand they convey the order of the world (space–time), while on the other hand they lay hold of that concrete (practical and social) space–time wherein symbolisms hold sway.
>
> (Lefebvre 1991: 42)

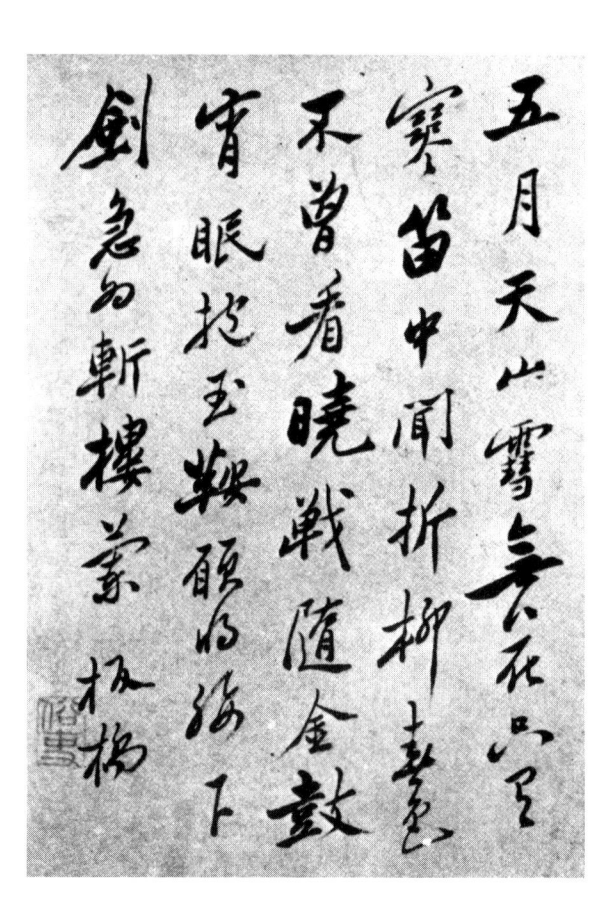

Figure 10.1 Calligraphy by Zheng Banqiao (1693–1765).

Lefebvre did not move on to observe Chinese urban space as he did to European cities. If we push this argument further, we arrive at this reading: Chinese cities do not have such a distinction but combine the two spatial orders, the conceived and the lived, where perspective and formal geometry do not dominate, but instead merge with a messy world of actual life and its religious and anthropological expression. If we turn to observe a real historical city in China, such as imperial Beijing (1420–1911), we do find a spatial pattern of this kind. It is different from that of a Renaissance European city. Without a bright and frontal order of public space dominated by the logic of visualization or perspective, that is, without an optical and public centre defined by high façades, squares and long, straight avenues, historical Beijing is a dense world of compounds, streets and laneways, in which an enclosed palace and hundreds of temples serving as local civic centres are scattered across the city and further into the rural areas, linked together with webs of streets. Yet the whole city is geometrically well-organized. A total rational order and the localized world of actual daily practices and religious life are combined. Without the telescopic perspective employed for open urban space, in the Chinese city there is no domination of the optical, visual and formal above the lived and the religious–anthropological as one finds in a Renaissance city (Figures 10.2, 10.3 and 10.4).

Figure 10.2 A map of Rome in the eighteenth century, showing streets opened up by the Popes since the fifteenth and sixteenth centuries.

Figure 10.3 View of an Ideal City by an artist in the School of Piero della Francesca, c.1470.

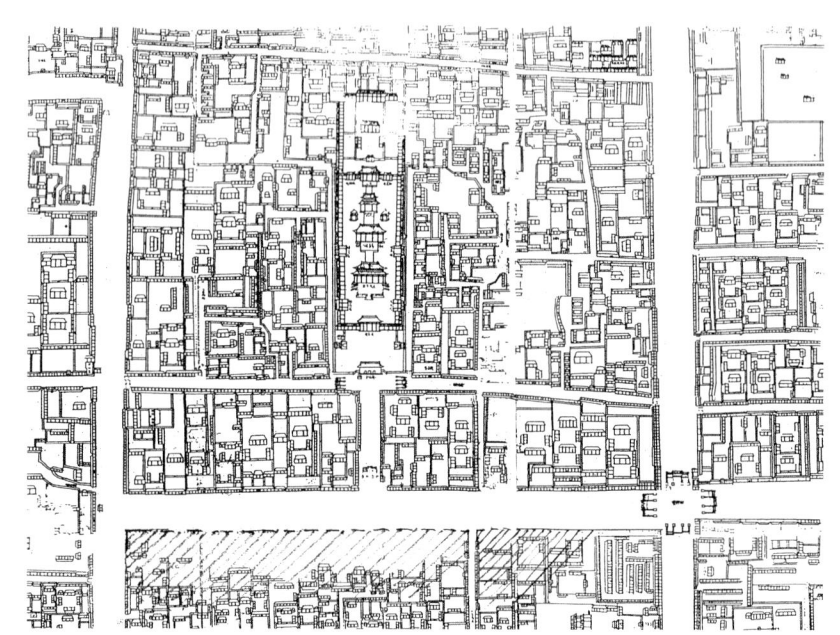

Figure 10.4 A map of Longfu Si Temple and the urban texture around, Beijing, 1750.

There are two important issues in such a comparison between Europe and China. The first is the way contraries or binary poles ('conceived' and 'lived') are dealt with in the two traditions. The second is urban layout as a reflection of a worldview or epistemological framework in the two traditions, which is to say, the city as a reflection of how its inhabitants see the world in the tradition in which the city was formed.

In his comparison of Chinese and Graeco-European philosophical traditions, Francois Jullien makes an extensive study of the approach to binary poles or contraries in Aristotle and in ancient China (Jullien 1995: 249–58). Jullien argues that both cultures have identified contraries as the base for change, yet the two traditions deal with the inter-relation between the contraries differently. While the Greek conceives the binary poles as exclusive and confrontational (as formal categories), the Chinese regards the contraries as mutually related and transformative. Without formalizing the contraries, that is, seeing them as formal *and* material at once, *yin* and *yang* (or any binary poles) are not mutually exclusive but interact with each other ceaselessly. This relational approach to contrariety in China explains various aspects of sociocultural practice in which China often adopts a middle path between conceptual poles identified in the European terminology, from the abstract conceived–lived contrast in the past to the communist–capitalist categories at present. Indeed, Chinese characters, or square words, are abstract signs and concrete images; it is a conceptual writing system and a lived enactment of the body and person in writing. The urban layout in historical Beijing also includes both a geometrical regularity and a scattered life-world of religious and anthropological expressions.

Regarding the issue of city as an epistemological layout, Jullien's study is illuminating as well. He employs an analogy between knowing to seeing to extend epistemological effort to visual experience in painting. According to him, while the European approach to knowledge includes a panorama of things, a top–down gaze upon a world that can be organized homogeneously, the Chinese approach to visual knowledge involves, not an overall perspective, but a series of views from one situation to another, as on a journey through which relations between views may be sketched in, like a meandering path in the mountains in a Chinese landscape painting (Jullien 1995: 123–4). We may in turn extend this analogy to our experience of urbanity as these two images of an epistemological approach in fact also correspond to the urban layout in a Renaissance city and a historical Chinese city such as Beijing, respectively. If the first is constructed with the logic of the perspective aiming at comprehending a totality, the second is built with changing views of long and shifting journeys linking one locality to another. If the first is optical, formal, centric, with the logic of the gaze, the second is invisible (always 'behind' and 'around the corner'), localized, corporeal and experiential. If the first is immediately grand and heroic, the second is visually humble and horizontal, small in any locality yet large in an endless unfolding of the local.

This brings us to Roland Barthes' reading of Tokyo, again with a comparative reference to Western cities. Among his many observations on Tokyo in his *Empire of Signs*, two appear most relevant here. The first concerns its empty centre and the second its spatial logic that requires intense labour of a journey. For the first, Barthes suggests that all Western cities have a centre, a site of fullness and 'truth', for spirituality (churches), power (offices), money (banks), merchandise (department stores) and language (agoras, cafés and promenades) (Barthes 1970 [1982]: 30–2). In Tokyo, however, one encounters a paradox: the city has a centre but it is 'empty', for the centre of the city is an enclosed imperial palace that is invisible to the outside. City life takes permanent detours and returns around a centre, a subject, a truth that is 'empty' (Barthes 1970 [1982]: 32). Of the second, Barthes talks about difficulties in finding a place in Tokyo due to a numbering system for the postal address that is unclear or inconvenient to outsiders (Barthes 1970 [1982]: 33–6). This observation, however, concerns not just the postal address, but an entire spatial layout of a city that is localized and demands corporeal labour and experience in order to be read and understood:

> The city can be known only by an activity of an ethnographic kind: you must orient yourself in it not by book, by address, but by walking, by sight, by habit, by experience; here every discovery is intense and fragile.
> (Barthes 1970 [1982]: 36)

Historical Tokyo or Edo share comparable spatial qualities with southern Chinese capital cities such as Nanjing and Hangzhou. They all had enclosed imperial palaces. The geometrical forms of these enclosures were all irregular due to topographical conditions such as rivers, lakes and hills. The layout of the enclosed palace in Beijing (and in other northern capitals such as Changan and Bianliang) was more regular or geometrical in shape on flat land. Yet the topological pattern of an enclosed imperial palace with a dense street layout and active localized uses remain consistent between these northern and southern cities. Beijing, Edo and some other cities in the region are comparable in this sense. In fact, both of Roland Barthes' observations on Edo can be borrowed to describe historical Beijing. Imperial Beijing is a city with an 'empty centre', for the central areas are invisible or, more precisely, are only partially visible in the long walls, trees and city moats that surround the forbidden centre. Second, historical Beijing is visually forbidden or empty yet pervasive in a local dispersion: it is a city that can be discovered only through walking and localized experience that is 'intense and fragile'.

We have highlighted the local, corporeal, experiential characteristics of historical Chinese cities such as Beijing. In relation to a Chinese approach to contrariety and the conceived-and-lived quality of Chinese writing identified above, we have also realized that there is a synthetic or relational middle path in the Chinese tradition in the city in which a conceptual geometry and a lived space of concrete life, corporeal experience and religious–anthropological expressions are closely combined. It seems that, compared with the European Renaissance tradition, a sphere of the lived, local and informal is more closely internalized with the conceived, total and formal in Chinese cities so that the overall urbanity reveals more clearly an organic world of liveliness and formlessness.

It is important to note that in Lefebvre's theory, lived space, being related to a life-world of energies and desires with religious and artistic expressions, contains revolutionary potential that can be employed to challenge a conceived world of abstractions and the rationalities of power and knowledge in the modern state, market capitalism and technological systems. Lefebvre's Marxist and postmodern agenda against instrumental modernity is clear in his own writings. To relate his ideas to the tradition in China, however, should not lead us to conclude that there is anything leftist or revolutionary in the Chinese tradition itself. The relevance in my borrowing his ideas here lies in the possible potential of a tradition as a critical resource to correct a one-sided modern development. The relevant potency of the Chinese tradition here is a balanced, synthetic, relational approach that may help to correct the abstract and instrumentalist in modernization.

For the task of reconstituting a humane and habitable urbanity for the public today, the structural tolerance characteristic of the lived, local and informal in the Chinese urban tradition seems particularly important. For a closer reading on historical Beijing, four observations can be made.[1] Three concern the spatial qualities of the city, and the last the Chinese revolution that conditions our learning from the past today: nature, scale, formlessness and historical gaps and continuity.

NATURE IN HISTORICAL BEIJING

In historical Beijing, the built fabric and the natural landscape were closely inter-related. Urban and rural areas co-existed inside the city wall (especially in the Outer or Southern City); urban functions extended outside the city walls, especially the numerous temples scattered in the hills and mountains around Beijing; and large areas of natural landscape (lakes and hills) were to be found inside the Imperial and Palace City at the centre of Beijing (Figures 10.5 and 10.6). In this last instance, we also witness a co-existence of a regular geometry of built fabric (walls and courtyards of the palaces) with an irregular geometry of nature, especially the six lakes meandering through the city. Without conceptions presenting a dualistic confrontation between humans and nature, the Chinese have developed one of the longest histories of appreciating nature and landscape in the world. (Berque 1993: 33, 34; 1997: 1)[2] This is manifested not only in the synthesis of contraries in the urban landscape (urban–rural, natural–artificial, regular–irregular), but also in other forms of cultural expression (poetry and painting) and in philosophical discourse as well (Taoism, Confucianism and yin–yang cosmology).

Figure 10.5 A map of Jingshan Hill on the central axis, part of a series of imperial gardens in central Beijing, 1750.

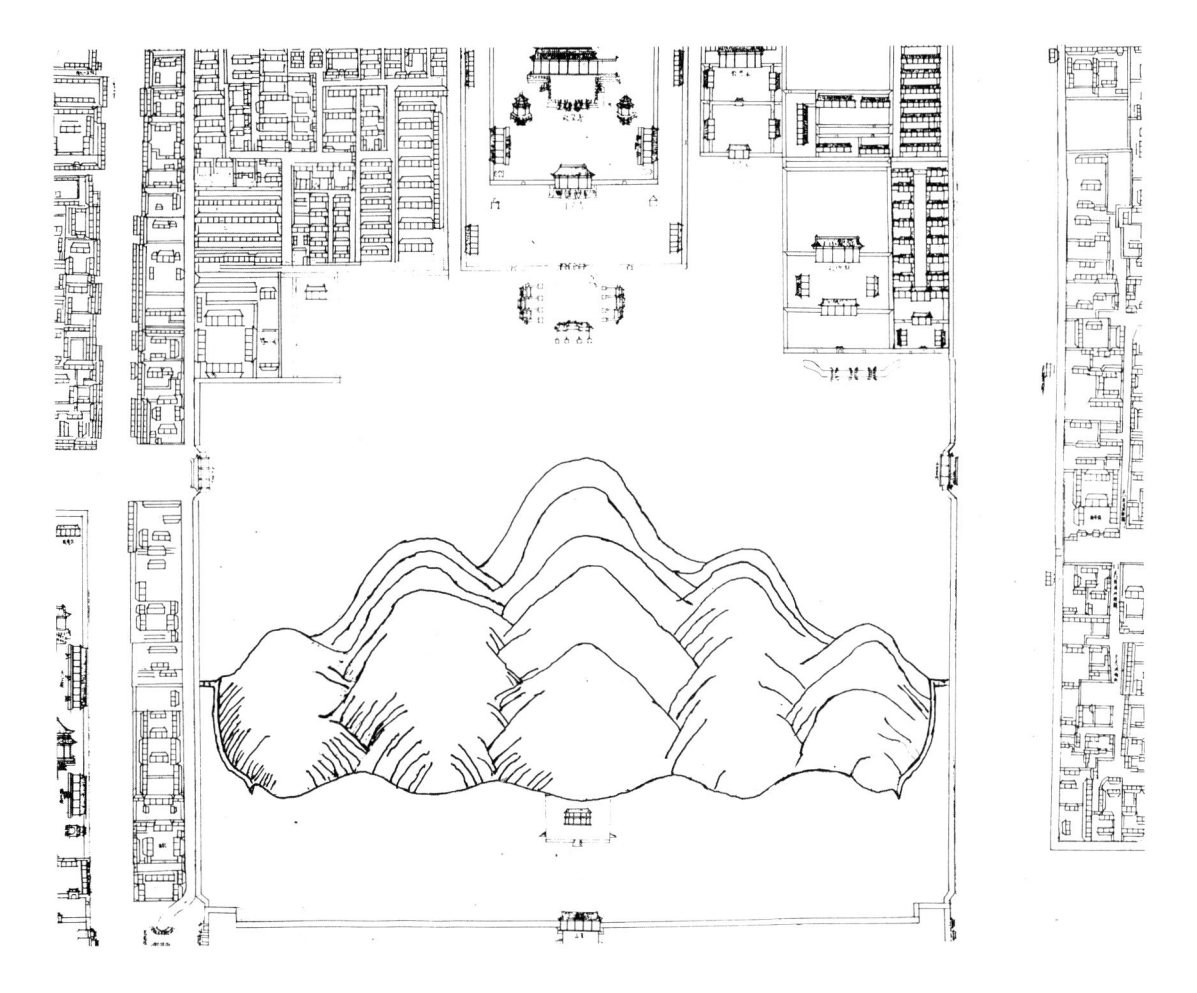

Figure 10.6 Jingshan Hill with a pavilion at the top, and the pagoda of the 'Northern Sea' (Beihai Park) behind, viewed from the east in central Beijing in the 1990s.

For the problems confronting us today, what can be learned here includes an ecological ethics in the layout of a human habitat in which nature takes primacy and human agency is considered relational with respect to nature, and a specific pattern of eco-urbanity in which landscape, rural conditions, the juxtaposition of nature and artificial urbanity, and an intermingling of regular and figurative geometries can be actively utilized.[3]

SCALE AND QUANTITY IN HISTORICAL BEIJING

In the ancient city of Beijing, compositions co-existed at different scales, as is most clearly evidenced in and around the Imperial Palace where large-scale courtyards co-existed with compounds of smaller scales. At the largest scale, the whole of Beijing was strictly organized with axes and references to the four cardinal points; at the smallest scale, minute houses, courts and gardens were also well-regulated with compositions relating to these axes and primary orientations. According to Wang Guixiang and Wang Qiheng, Chinese teaching of design emphasizes the co-existence of *dazhuang* (the large and sublime) and *shixing* (the appropriate), that is, a large composition and adequate local forms and spaces (Wang, G. 1998: 477–91; Wang, Q. 1992: 117–37). The idea is also expressed in the word *xing* (shape or form of a smaller scale) and *shi* (tendency and propensity of a larger scale). For example, a well-known Chinese teaching says 'consider forms at a distance of one hundred feet, and a dynamic lifeline at a distance of one thousand feet' (*baichi weixing, qianchi weishi*). This co-existence of compositions at different scales can be considered a case of fractal geometry with self-similar patterns recurring at all scales, as we find in clouds and plants (Mandelbrot 1982). The design in Beijing intuitively implements this natural or organic logic into an artificial and geometric urban-architectural construction (Figures 10.7, 10.8 and 10.9). While fractal geometry captures natural forces of growth and additions of quantity, Euclidean geometry concerns instead pure and static forms and proportions (Mandelbrot 1982: 1–13). While Euclidean or Greek geometry has focused on pure form and formal relations, ancient Chinese studies have emphasized geometric shapes containing numerical or quantitative problems (see Engelfriet 1998: 106–7).[4] In other words, the 'fractal geometry' of imperial Beijing is not a geometry of form and proportion, but one of number and quantity. It includes a compositional logic of growth, addition and proliferation, the spatial logic of the 'more'.

The lesson here includes recognizing a non-formal geometry of quantity and substance that permits external additions and internal intensifications. A concrete message is that a mega-building or large urban structure is not a final condition, but an initial stage to which external additions of equal or larger frames and internal, smaller frames and spaces can be included. When forms of 'one hundred feet' and a dynamic propensity of 'one thousand feet' are all included, a rich human construction can be expected which has not only large-scale accommodative capacities, but also micro-spaces at a local, human scale. Large and small are not contradictory but can be well-related, each playing its own role.

DISSOLVING FORMS IN HISTORICAL BEIJING

In historical Beijing, one will not find the centre of 'truth' that Roland Barthes identifies in Western cities, or the conceptual rational perspective space that Henri Lefebvre finds in European Renaissance cities. The traditional city of

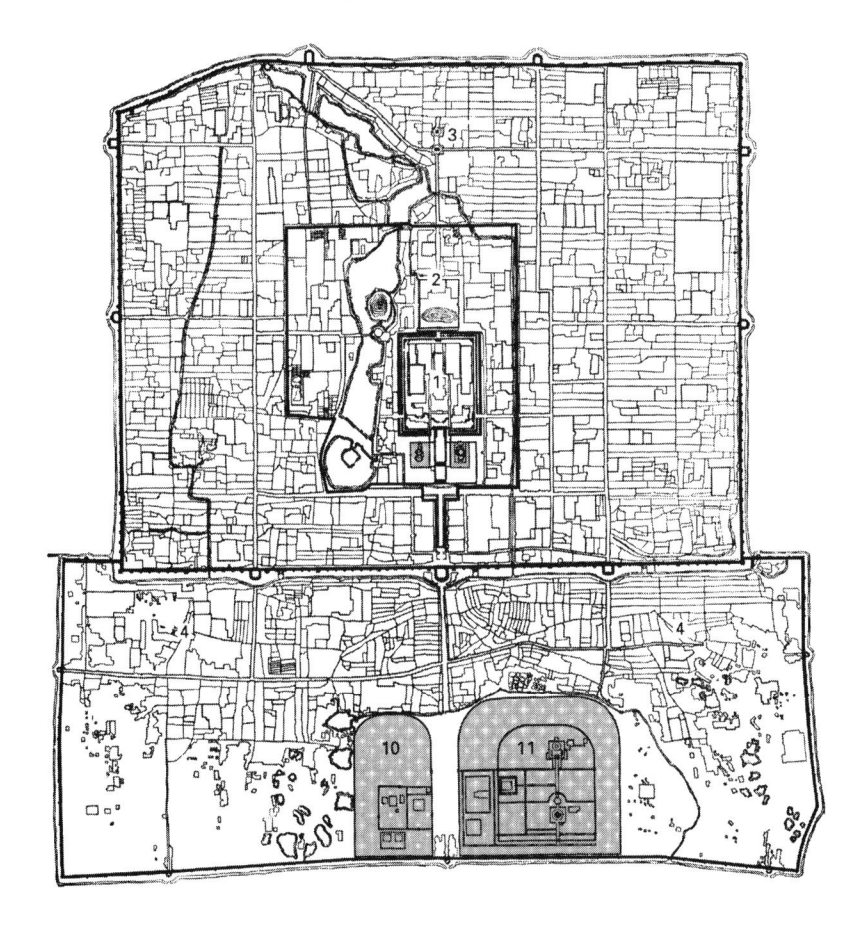

Figure 10.7 A map of imperial Beijing, 1553–1911.

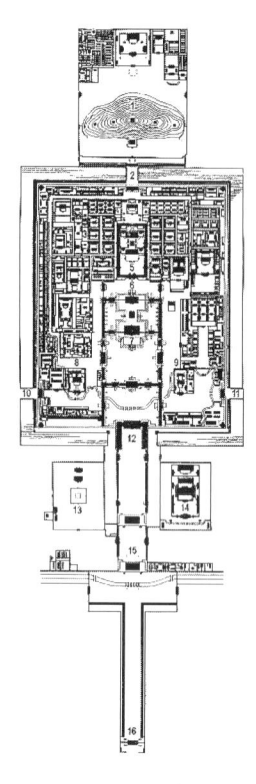

Figure 10.8 A map of central Beijing showing the Forbidden City and the approach in the south and the Jingshan Hill in the north, 1644–1911.

Figure 10.9 An aerial view of central Beijing from the north in the early twentieth century.

Beijing unfolded with a different spatial logic. It had an overall geometry of the entire city rigorously implemented, yet it also had a pervasively localized space of life-worlds of social communities with religious and anthropol. In this extensive city of life experience, scattered local centres such as temples (Figure 10.10) were interconnected with webs of streets and laneways that channelled the endless journeys leading to these centres. This composition included a geometry of magnitude and quantity. It also included a dispersed texture or field of local societies and communities, with their daily life and festive expressions unfolding around temples, teahouses, restaurants, guildhouses and various shops along streets, in a constant and flexible use of space by the population (Figure 10.11).

Built forms were not solid and distinctive, as was the tall, stone architecture of European cities, but were low-rise, soft and layered (with timber, stone, brick, tiles, bamboo and wooden fences, curtains of signage and so on) with shifting social activities (commercial, religious and everyday). Form, as a pure, clean and distinctive shape of objects as one finds in the European cities

THEORETICAL LEGACIES OF HISTORICAL BEIJING

was dissolved. This was achieved by a domination of small and grouped horizontal structures and the enclosure of significant buildings like palaces, government offices and temples by courtyard walls, together with the soft surfaces of built forms along public/commercial streets. It was helped by the weak or multiple programming of spaces so that a temple was also a commercial focus, accommodating monthly or yearly festivals and other functions. Pure form was dissolved further by the constant and flexible use of space by the population, turning a house compound into a temple, and permitting a teahouse or restaurant to accommodate a theatre, not to speak of the more amorphous uses of streets and other outdoor spaces.

Figure 10.10 Temples as centres of local society scattered in and beyond the city of Beijing, 1644–1911.

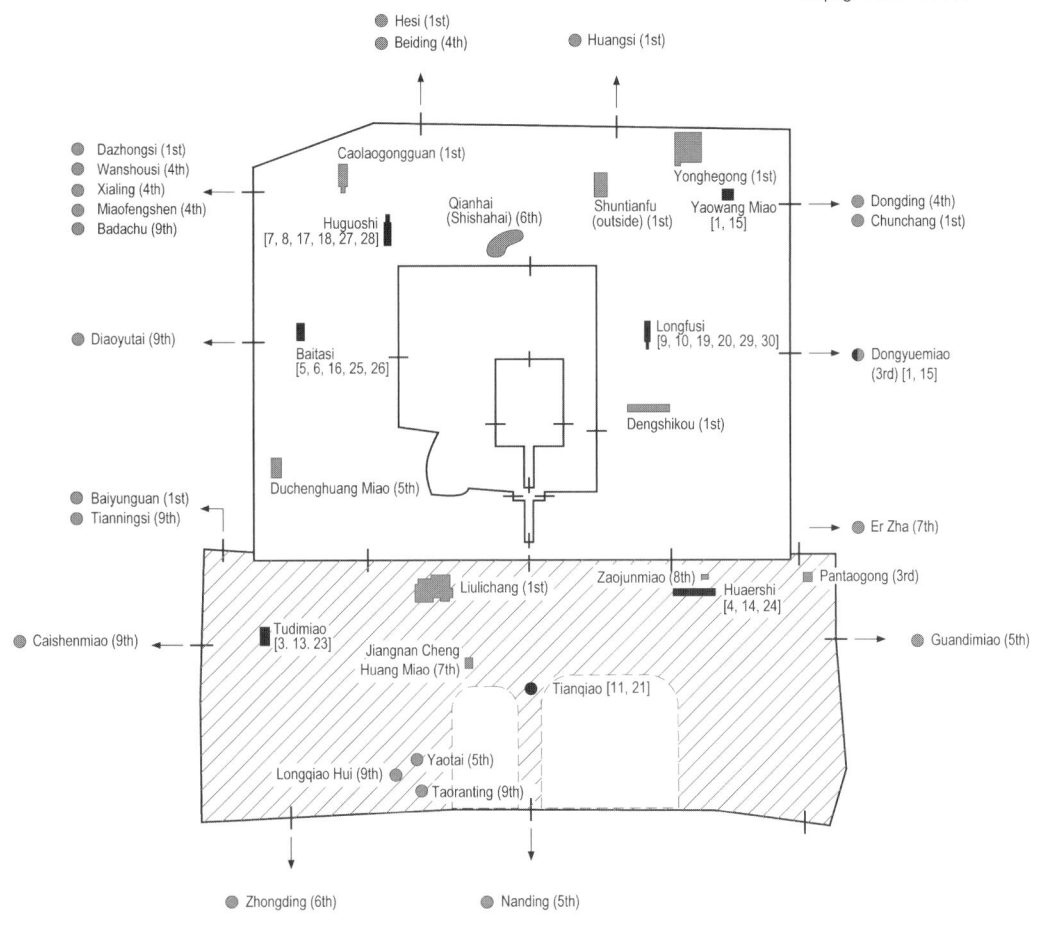

KEY
- Temple and fair sites opened monthly [with the date of the fairs]
- Temple, fair and pilgrim sites opened for annual worship (with the month in which they were visited).
- Areas of concentration of guild halls, theatres and theatre-restaurants.

135

Figure 10.11 Street scenes of Bianliang (Kaifeng), the capital of Northern Song dynasty (960–1126): sections of a scroll painting *Qingming Shanghe Tu* by Zhang Zeduan (active 1101–1124).

At a socio-political level, Beijing also included a relational synthesis of a total geometry of the formal state and a dense texture of local informal society with its transactions. Sociologists such as Philip C. C. Huang have uncovered the informal, relational mediation between the formal organization of the state and informal local communities (which resolved two-thirds of all legal cases in one empirical study of late Qing dynasty) (Huang 1993: 216–40). In this situation, moral and political order was secured neither by a tyrannical state nor a critical civil society in confrontation, but through relational mediation between state and society, and between formal regulations and informal socio-personal interactions within a shared culture.[5]

There are several aspects of this tradition we can adopt for today's design practice and social critique. To accommodate the lived or a dynamic lifeworld of social communities, urban and architectural design can be conceived as constructing a field or texture with scattered local centres or intensities rather than a formal object. By adopting a fractal and formless geometry of quantity, energies and intensities can be added internally or externally in an extensive field.[6] Weak and multiple programming can contribute to a liberal accommodation to the informal use of space and its underlying forces and potentials. Such approaches to designing the informal, using a formless and quantitative geometry, may contribute to a new political ethics which does not aspire to a confrontational critique of state by the society, but sponsors an ongoing relational critique in which a progressive agenda is implemented in an embedded practice involving many different agents and forces.

THE CHINESE REVOLUTION

The above three observations on historical Beijing and the lessons we identified require a qualification if they are to be useful today. Historical and modern Beijing, or imperial and modern China, cannot be simply equated: the revolution and modernization that have occurred separate us today from the past. Regarding modernization in the urban world, the scale and fabric of the city today have been much altered. Constructions are larger in scale and more vertical, that is, they are more three-dimensional and object-oriented; and the fabric has been torn apart and remade with formal geometries of a scale even larger than that in the Renaissance and nineteenth-century European cities. The past three decades have witnessed an explosion in size of this city, that started in the early modern (1910–30s) and socialist period (1950–70s). The ideas learned from historical Beijing therefore have to be distilled to deal with the problems we are facing today. The idea of a geometry of substance or quantity, of extensive texture, of local intensities, of informal use and liberal programming, may be especially useful today to rehumanize the formal, object-oriented geometries that dominate contemporary Beijing.

Regarding the Chinese revolution, we must understand that radical and profound changes have occurred separating the present from the past. If we examine these revolutions closely, we realize that a central theme and a major consequence, regardless of ideological positions and despite the turbulent changes, is the rise of the public in modern China. There was already a public when the modern press and urban facilities were developed in the 1910–30s. Yet this public cannot sustain itself without internal support and protection of a strong sovereign nation-state against foreign domination. After 1949 a public was established, but it was collectivist and regimented. The radical left from 1957 to 1976 brought this collectivist uniformity into chaos. Today the public is mutating into a liberal and individualized society, yet Western types of checks and balances through a transparent press and an independent legal system are still missing. A liberal horizontal civil society is burgeoning, yet the critical vertical mechanism linking it to the state is lagging behind (if such a transformation is to happen at all). What is clear so far is that the 'people's democracy' that Mao Zedong and the Communist Party had been promoting since 1940 is not radically different from the 'bourgeois democracy' that the Nationalist Party has been exploring and experimenting with (Mao Zedong 1966–70: 623–70). In historical hindsight, and as evidenced now in both Taiwan and mainland China, the left and right of the Chinese revolution are converging in a common liberalization of the public.

Regarding the shape of a liberal and democratic public, the central question remains as to the form or institution that this public in mainland China is to take. Will a free press, an independent legal system and multi-party elections, or certain aspects of these practices, ever happen one day in China? Given the path already travelled by the Chinese in the twentieth century, political institutions with Western sources are likely to be adopted further, with Chinese modifications. What is likely to emerge in China is a combination of liberal and democratic institutions with relational and informal practice. In the recent debate on civil society and the public realm in relation to China, scholars highlighted the specificity of the Western model and the relational ethics in China's past in which a powerful 'third realm' asserted a critical role between state and society (see Brook and Frolic 1997: 3–16; Fei 1948 [1992]: 69–70; Huang 1993: 216–40). With the weight of tradition, it would be impractical not to expect this Chinese ethics to re-emerge and assert its capacities for a harmonious and functioning society. There is a tendency towards social democracy rather than a 'strong' capitalist liberalism in contemporary China. Within the Communist Party, there are voices against the Leninist legacy; and there are debates that reveal a preference for a European-style social democracy as an alternative to Soviet communism and American capitalism (Xie 2007: 1–8).[7] From the outside, under its new leadership the Party has been active in 2006 and 2007 rebuilding public goods (including clean environment and social services) and pursuing social equity and justice in an effort to contain market aggression and single-minded developmentalism, while leading and managing sustained development at the same time. We may conclude that it is now historically possible, and politically progressive, to envisage a new public domain with a relational ethics and an organic urbanity, in the Chinese tradition of the synthetic that cultivates the local and the informal, in a city of nature, quantity and formlessness.

NOTES

1 The following observations on historical Beijing are empirically based on my earlier work (Zhu 2004), especially chapters 3 and 9.

2   According to cultural geographer and theorist Augustin Berque, 'at the scale of world history, the notion of landscape was first invented in Southern China towards the beginning of the fifth century' (Berque 1993: 33, 34; 1997: 1).
3   The debate on the concept of a mountain–water city in the late 1990s in China is an indication of renewed interest in the Chinese tradition of a landscape urbanity (see Bao and Gu 1996).
4   I have touched on this issue, in chapter 2 of my book, *Architecture of Modern China: A Historical Critique* (2008) in the context of Greek geometry being introduced into China in 1607 and the gradual rise of formal geometry of the object in modern China.
5   One of the earliest modern studies to identify a relational and mediating relationship between state and society in the Chinese tradition is Fei's *Xiangtu Zhongguo* (1948: 22–30), or, in English translation, Fei (1992: 69–70).
6   In Cecil Balmond's exposition of an 'informal' approach to the design of building structures, the local or local intensities are also considered a crucial generative moment for a non-Cartesian structure and, if I may add, a non-Euclidian or non-formal geometry (see Balmond 2002: 217–28).
7   The most eminent voice today that has criticized the Leninist tradition and supported European-style social democracy as alternative to Soviet communism and American capitalism is Xie Tao in his '*Minzhu shehui zhuyi moshi yu zhongguo qiantu*', 2007: 1–8, published in *Yanhuang Chunqiu*, a Beijing-based monthly with a large circulation among Party members and high-ranking officials. For a report on this see Jiang (2007a: 30; 2007b: 32–4).

BIBLIOGRAPHY

Balmond, C. (2002) *Informal*, Munich: Prestel.
Bao, S. and Gu, M. (eds) (1996) *Chengshi Xue yu Shanshui Chengshi* [Urban studies and a 'mountain–water' urbanism], Beijing: Zhongguo Jianzhu Gongye Chubanshe.
Barthes, R. (1970 [1982]) *Empire of Signs*, New York: Hill & Wang.
Berque, A. (1993) 'Beyond the modern landscape', *AA Files*, 25: 33–7.
—— (1997) 'Landscape in Japan as a symbolic form', unpublished paper presented at a research seminar in Melbourne.
Brook, T. and Frolic, B.M. (1997) 'The ambiguous challenge of civil society', in Brook, T. and Frolic, B.M. (eds) *Civil Society in China*, New York: M. E. Sharpe.
Engelfriet, P.M. (1998) *Euclid in China: the Genesis of the First Chinese Translation of Euclid's Elements Books I–VI (Jihe yuanben; Beijing, 1607) and its Receptions up to 1723*, Leiden, Boston and Koln: Brill.
Fei, X. (1948) *Xiangtu Zhongguo* [Rural China], Shanghai: Guanchashe; transl. 1992 by Hamilton, G.G. and Zheng, W. as *From the Soil: the Foundations of Chinese Society*, Berkeley, CA: University of California Press.
Huang, P.C.C. (1993) '"Public sphere"/"civil society" in China: the third realm between state and society', *Modern China* 19, 2: 216–40.
Jiang, J. (2007a) 'Minzhu shehui zhuyi de tupo' [A breakthrough for social democracy], *Yazhou Zhoukan*, 25 March: 30.
—— (2007b) 'Zhonggong yuanlao zhichi huwen zhenggai' [Veterans of the Chinese Communist Party support Hu and Wen's political reform], *Yazhou Zhoukan*, 10 June: 32–4.

Jullien, F. (1995) *The Propensity of Things: Towards a History of Efficacy in China*, J. Lloyd, trans., New York: Zoon Books.
Lefebvre, H. (1991) *The Production of Space*, D. Nicholson-Smith, trans., Oxford: Blackwell.
Mandelbrot, B.B. (1982) *The Fractal Geometry of Nature*, San Francisco, CA: W.H. Freeman.
Mao Zedong (1966–70) 'Xin minzhu zhuyi lun' [On new democracy], in Zedong, M. *Mao Zedong Xuanji* [Selected works of Mao Zedong], vol. 2, Beijing: Renmin Chubanshe: 623–70.
Wang, G. (1998) *Dongxifangde Jianzhu Kongjian: wenhua kongjian tushi ji lishi jianzhu kongjian lun* [Architectural space in the East and West: a theory of space in cultural schema and historical construction], Beijing: Zhongguo Jianzhu Gongye Chubanshe.
Wang, Q. (1992) *Fengshui Lilun Yanjiu* [Studies in the theory of Fengshui], Tianjin: Tianjin Daxue Chubanshe.
Xie, T. (2007) 'Minzhu shehui zhuyi moshi yu zhongguo qiantu' [Social democracy and China's future], *Yanhuang Chunqiu* 2: 1–8.
Zhu, J. (2004) *Chinese Spatial Strategies: Imperial Beijing 1420–1911*, London: RoutledgeCurzon.
—— (2008) *Architecture of Modern China: A Historical Critique*, London: Routledge.

11 ECO-CITY? ECO-URBANITY?

ARVIND KRISHAN

INTRODUCTION

A globally accepted definition of sustainable development is that it 'meets the needs of the present without compromising the ability of future generations to meet their own needs' (Brundtland Commission 1987). We share this beautiful planet with blue skies and seas with 70 million other species in an eco-system both dynamic and finite, driven by the regenerative resources of the sun. The challenge for contemporary designers is to use these regenerative powers to design a sustainable habitat within the carrying capacity of the finite eco-system that is our Mother Earth.

To design our cities and buildings sustainably we must recognize the limited carrying capacity of the environment and produce integrative and inclusive solutions rather than mutually exclusive ones, as is the current trend. This is required if we aim to achieve economic growth in line with, rather than at the expense of, environmental and cultural diversity and quality. Sustainable design must integrate considerations of resource and energy efficiency, create healthy buildings and materials, use land in ecologically and socially sensitive ways and develop an aesthetic sensitivity that inspires, affirms and ennobles.

THEORETICAL FRAMEWORK OF ECO-SETTLEMENTS

An eco-settlement is based on ecological planning and design principles. Eco-settlements are those that promise to be sustainable in the long run, comprising human habitation systems that are environmentally benign, socially

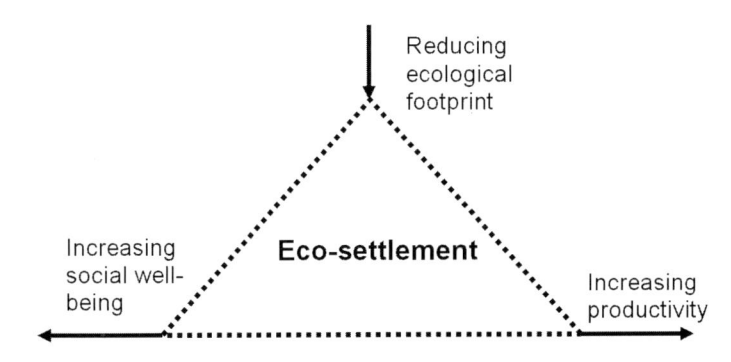

Figure 11.1 Conceptual framework for eco-settlement.

equitable, culturally appropriate and economically progressive. While it may be naive to assume that these four conditions can ever co-exist, we can and should attempt to arrive as close as possible to this ideal.

A holistic approach to sustainable eco-settlements

When we discuss sustainable urbanism and architecture we usually refer to the principles of ecological sustainability. Translating these principles into action is needed for not only the built environment, but the planet itself. In these times of unbridled globalization in which cultural integrity is constantly under attack, we need to focus on cultural sustainability as well as responsible and accountable development. This calls for sensitivity to the impact of global ideas, practices and technologies on local social and cultural practices (Radović 2007).

Environmental, economic and social sustainability are interdependent and interlinked (Figure 11.3). While the physical dimension of built form dominates our perceptions, it cannot be disassociated from its economic and environmental context. While we can develop and fine-tune planning and design processes, sustainability is intrinsically linked to issues of local social equity.

Tradition as a repository of knowledge

Tradition is not merely the romantic association of a group of people with their past. Planning and development processes often fossilize tradition and reduce it to a mere tourist attraction. However, sensitivity to tradition allows us to excavate the sophisticated repository of knowledge embedded in planning and design principles and processes linked to the ecological and socio-

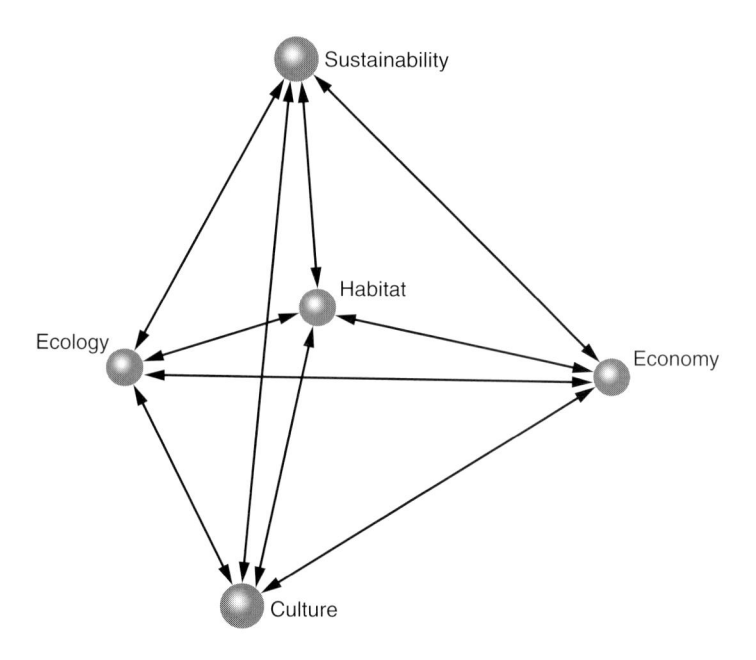

Figure 11.2 Holistic approach.

Environmental sustainability
Ecosystem integrity
Carrying capacity
Biodiversity

Human well-being

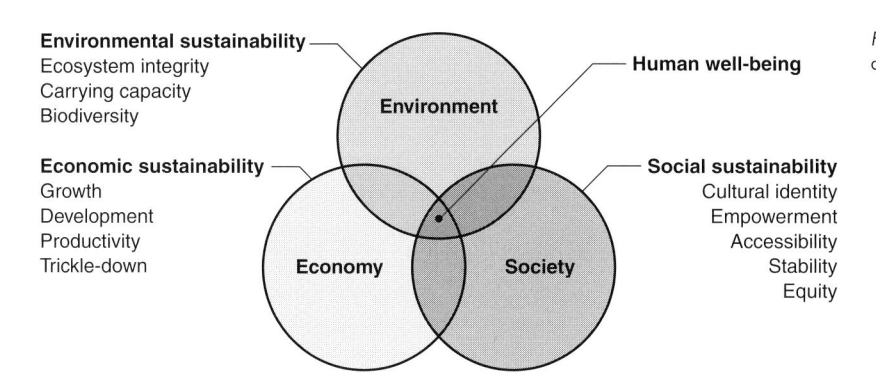

Economic sustainability
Growth
Development
Productivity
Trickle-down

Social sustainability
Cultural identity
Empowerment
Accessibility
Stability
Equity

Figure 11.3 Human well-being and related elements.

economic contexts of times past. These repositories can unearth the knowledge base that we need in our current threatened eco-system. While conventional habitats and buildings waste natural and man-made resources, sustainable built form integrates all resources that are a generic part of habitats.

WHAT FORM OF HABITAT?

Trends indicate that the growth of settlements and their outward spread will continue eating into land required for agriculture, forest cover and even wetlands, and by 2051 an ever-decreasing amount of agricultural land will have to cater for nearly double the amount of food that will be needed. For example, it is forecast that in India by 2051 ten million hectares of good agricultural land will be consumed by the expansion of settlements, most of which will spread both vertically and horizontally. India and China are projected to be home to 40 per cent of a total world population of ten billion by 2030. Urbanization in both countries is projected to reach 75–80 per cent, meaning that 3.5 billion people, or half the total current world population, will live in Indian and Chinese cities. In face of this relentless urbanization it is prudent to re-examine the critical issue: what form must this habitat take? The best answer is a 'walking city', an idyllic city where the fundamental relationships of living and working places are entwined with nature – the environment.

The motor car has dominated urban form for the last century, resulting in a polluted, environmentally stressful and resource-guzzling habitat that is ubiquitously perpetuated and has become the model for all development. We must stop perpetuating this urban trap for future generations.

ENERGY RESOURCE FLOW: AN ECOLOGICAL FOOTPRINT (EFP) MODEL

We need to address the issue of the projected growths and urban spread by using sound ecological models. An energy–resource flow EFP model seeks to answer questions of planning and sustainable design. The EFP model relates the inputs and outputs of the habitat. It models natural resources such as minerals, metallic products, organic and inorganic chemicals and forest products, primary energy used for heating or cooling, lighting and transportation of goods and people. It plots the natural resources of land, water, air and flora and fauna. The volume of input and the functioning of the built environment dictate the volume of emissions and wastes that are produced. The model shows how this waste could be recycled and used as an additional resource, thereby minimizing primary inputs and closing the cycle, rather than extending it in an open-ended linear form.

BUILDINGS AS CATALYSTS FOR SUSTAINABILITY

In the face of environmental crisis and climate change we are used to doomsday scenarios. But the environmental crisis can be converted into an opportunity. As planners and architects we can use it to develop habitat and architectural typologies leading to a new environmentally and culturally appropriate era in architecture.

Climate-responsive design

Every culture has evolved its built form in space and time in response to its ecological/climatic context, generating in the process a rich, culturally determined diversity in built form. This process can be analysed and scientifically developed. An attempt in this direction is illustrated in Figure 11.4. The process of design is in essence a process of design decision-making, as shown in Figure 11.5.

Micro-climates are small-scale climatic patterns resulting from such influences as topography, urban forms, water bodies and vegetation. A microclimate consists of any local deviation from the climate of the large region or

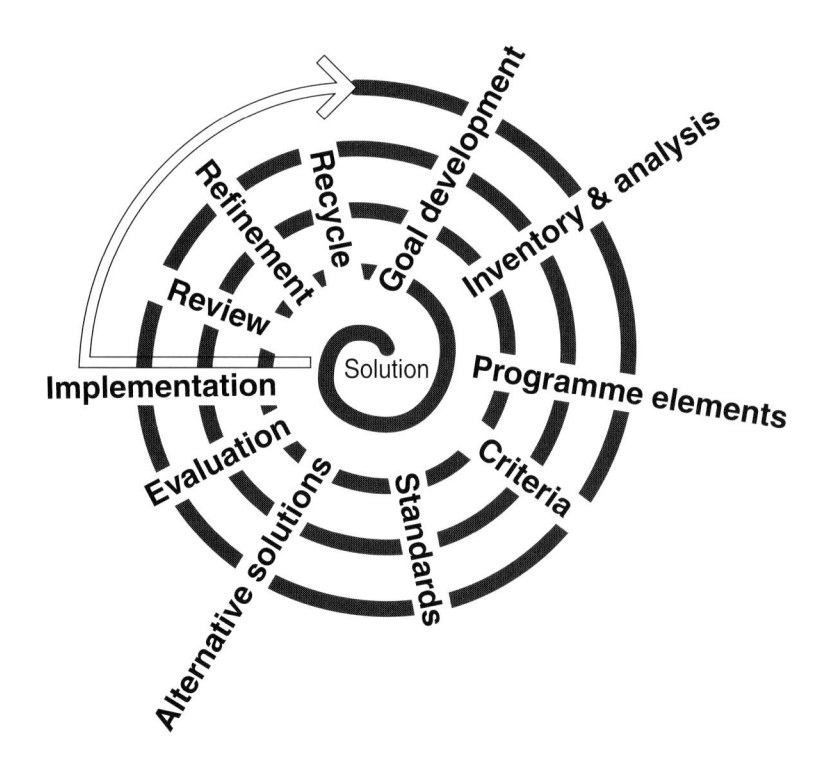

Figure 11.4 Graphical representation of process of design.

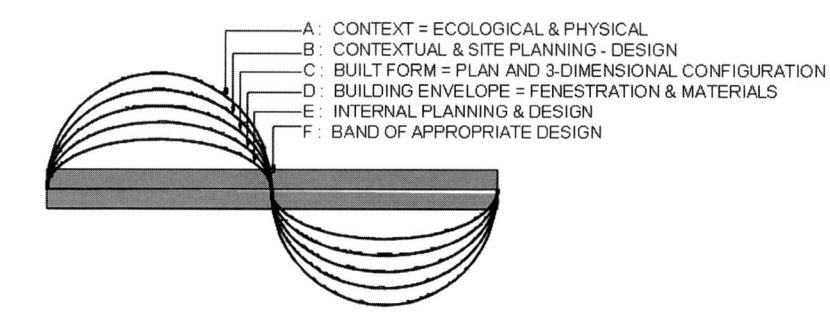

```
A : CONTEXT = ECOLOGICAL & PHYSICAL
B : CONTEXTUAL & SITE PLANNING - DESIGN
C : BUILT FORM = PLAN AND 3-DIMENSIONAL CONFIGURATION
D : BUILDING ENVELOPE = FENESTRATION & MATERIALS
E : INTERNAL PLANNING & DESIGN
F : BAND OF APPROPRIATE DESIGN
```

Figure 11.5 Ecological process of design.

zone in which it is found (Krishan *et al*. 1995; Radović 2007). A climate-responsive design takes account of the microclimate to achieve a sustainable building envelope. It attends to both the specific regional macro-climate and microclimate of the building and has a crucial effect on the energy resource demand to control the internal climate of a building. Figure 11.4 illustrates the sequence of ecological processes of the design and Figure 11.5 shows the integrated process of design and various tools of analysis developed by the author.

SOME CONTEMPORARY SOLUTIONS

The single parameter that embodies the state and use of various forms of natural resources is energy. In this section I discuss some contemporary architectural solutions that optimize energy use and resource consumption as well as critical embodiment of culture. These designs do not follow any preconceived notion of architecture. The projects express a symbiosis of ecological design and solar geometry. The design principles discussed below have emerged from my analysis of indigenous architecture and the scientific process of design, and have been translated into contemporary buildings at various places within different ecological contexts.

Building in a high-altitude, extreme cold and dry ecological context

Sustainable solutions for these projects incorporate the following elements and features achieved through architectural design a response to solar geometry, daylight and culture by using traditional construction practices. The result is a culturally appropriate form of sustainability that embodies tradition.

A hill council complex, Leh, India

This major civic structure designed and built as architectural design solution in this context is located at 3,514m above sea level in a cold, dry climate with a long, severe winter from October to the end of March (minimum dry-bulb temperature −30°C). The building configuration was developed in response to solar geometry by maximizing solar access for daylight and heat in this cold climate. A judicious use of thermal mass allows glare-free daylight to minimize the lighting load, and thus the consumption of energy in the building. Traditional local materials were used in the form of sun-dried mud blocks upgraded to stabilized mud blocks. This building achieves a new expression of Ladakhi architecture by harmonizing indigenous with modern forms and incorporating local materials. This helped to sustain artisan skills and the local art of construction, resulting in ecologically and culturally appropriate architecture (see Figures 11.6–11.7, colour insert 2).

Punjab Energy Development Agency office complex, Chandigarh

The Punjab Energy Development Agency (PEDA) office complex is an architectural response to a composite climate and its urban context in the city of Chandigarh. Located on a flat, practically square site with no major topographical variations, Chandigarh as a city lies on the plains at the foot of the Lower Himalayas in a composite climate context with extreme climate swings over the year. A very hot and dry period of almost four months (maximum dry-bulb temperature 44°C) is followed by a hot, humid monsoon period (maximum dry-bulb temperature 38°C and maximum R.H. 90 per cent) of about four months with intervening milder periods, then by a quite cold period of shorter duration (minimum dry-bulb temperature 3°C). Equally important for Chandigarh is its context in space and time. This bold experiment in city planning and architecture was based on the explicit design principle: to build with the

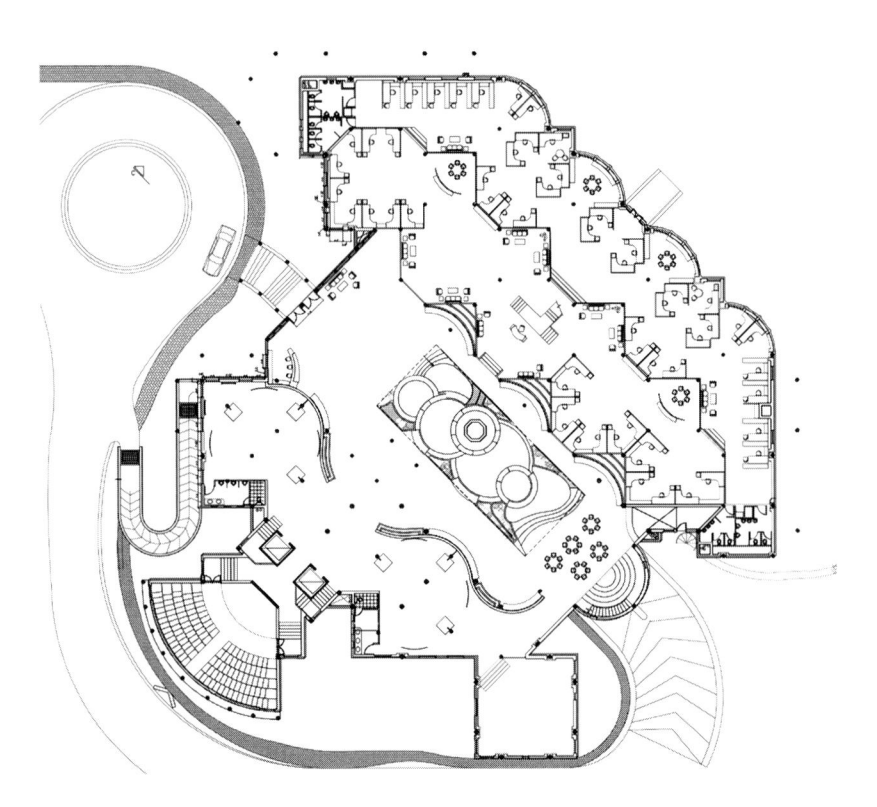

Figure 11.9 Plan of the complex.

climate. A holistic office design has been developed in which the floor plates float in a large volume of air and the building envelope interacts with the external conditions, rather than stacking the floor plates one on top of the other and segregating each floor plate as is conventional. The building configuration has been generated in response to the solar geometry rather than any preconceived design. Thus a unique built form has been generated in response to summer and winter requirements.

The innovative fenestration design consists of spherical solar shells that respond to the solar geometry. The natural movement of air is achieved through a down-draft wind tower coupled with vertical cut outs running through various floors and integrated with the solar shells via solar chimneys. A well-designed water body in the centre of the building provides a cooling element and restful environment. The atrium roof is designed with hyperbolic prefabricated units to respond to the solar geometry, allowing winter solar access and cutting the penetration of heat in the summer, permitting very good daylight distribution. The design is responsive to its climatic context and good daylight distribution, optimizing the consumption of electrical energy resulting in an architecture that responds to solar geometry, in which renewable energy systems such as photovoltaic panels are an integrated element of the roof design.

Sardar Swaran Singh: a new ecological architectural language for an innovative institute

The author was invited to participate in a limited competition to build Sardar Swaran Singh, the National Institute of Renewable Energy for the Ministry of Non-Conventional Energy Sources, Government of India. The innovative architectural design presented by the author was selected by the jury for the implementation of this project.

The strategies for planning and design included controlling the microclimate of the site by generating a water body drawn from the canal and by planting trees in the site. Each building in the entire complex was designed as a climate-responsive, solar-powered, passive-energy structure. The primary generator and tool for developing this low-energy building design was the architectural design, which sought to maximize environmental control through naturally conditioned laboratories and spaces and using natural daylight to reduce electrical consumption during the day. Evaporative cooling from the water body was integrated into the building design. The R&D wing was designed to connect naturally conditioned laboratories and spaces to the water body through wind towers and with earth tunnels to the laboratory and building spaces. A domed light vault designed to distribute adequate daylight was integrated with solar chimneys on the domed vault connected to wind towers.

An equally critical natural resource that is becoming more scarce is water. This project spreads over 34 ha of land on a highway with no municipal services or facilities. A complete scheme of water management includes direct recharge into the aquifer, collecting building run-off into the lake, recycling water from the root zone of the sewage treatment system and surface run-off seepage through ponds.

Himurja Building, Shimla, India: transforming the limitations of existing buildings into opportunities

Existing buildings consume disproportionate amounts of natural resources in the various forms of energy. By retrofitting them we can render these buildings energy efficient. One example of such a retrofit is Himurja Building in Shimla, a double-storied building previously used for shops and warehousing that was handed to us to be redesigned as a climate-responsive building and converted into offices.

The building was redesigned with four-and-a-half storeys, with a hybrid existing reinforced cement concrete frame and steel structure for the additional floors. The design optimizes the use of solar heat gain to all offices during critical periods of the year, ensuring adequate penetration of sunlight and protection from mutual shading. Glare-free daylight was used to minimize the lighting load and thereby the consumption of energy in the building. The distribution of heat in the building was achieved by using a double convection loop (see Figures 11.18 and 11.19, colour insert 2).

CASE STUDIES IN THE INDIAN CONTEXT

At the city scale the different urban patterns of Indian cities can be used as a laboratory for studying urban forms. A brief description of some of these patterns in very similar cultural and climatic contexts is presented to help us understand and develop strategies for habitat patterns that are ecologically and culturally appropriate. Shahjahanabad, New Delhi and Gurgaon are contiguous developments. Chandigarh, though 250 km away, shares the same cultural and climatic context.

The walking indigenous city: Shahjahanabad (Old Delhi)

The walled city of Shahjahanabad was built on the banks of River Yamuna, as a new capital city for 60,000 inhabitants. It consists of the king's palace and Chandni Chowk bazaar. Perpendicular to the palace is the main thoroughfare. Chandni Chowk St is 40 m wide and 1.5 km long. It was planned to be the backbone of the city and represents the most powerful structure of the city plan, housing 1,560 shops and porticos (Johnson 1969). Land use was zoned according to occupation and related activity patterns, with the tertiary sector located in the central core area and the services, shops and manufacturing units at the periphery. This arrangement took care of congestion-related problems as well as the appropriate location of land use according to their nature.

Shahjahanabad is a planned city. Even with its mixed land use, there was a certain hierarchy in the city plan. The Shah's palace dominated the urban form of the city, followed in importance by the place of worship, Jama Mosque. The third element of the city consisted of the dwellings and workplaces of the people located outside the palace but within the walled complex. The city was organized into *mohallas* (communities defined physically and socially within the same district) with *havelis* (private residences, a building typology indigenous to Indian culture in response to social and climatic context) and *katra* or courtyard houses forming typical modules. The typical building units were pressed back to back with narrow streets between them to provide access. These were perfectly functional for pedestrian traffic. Built to face extreme climate conditions, these buildings and streets protect residents from dust storms and cold winds in winter.

Before the 1930s people travelled in bullock carts or in horse-drawn carriages. Rickshaws, bicycles, *tongas* and trams came later in the 1940s. The only wide roads in Shahjahanabad were in Chandni Chowk, Hauz Qazi, Khari Baoli and Daryaganj. The small streets in front of houses, alleys and narrow lanes of the neighbourhood were as wide as was necessary before the arrival of the car. With its compact development, suitable for high-density core areas in which all activities occur at walking distance, Shahjahanabad takes up very little land. It permits limited access to traffic, especially motorized transport, since it was designed primarily for pedestrian traffic. The introduction of automobiles would lead to jammed traffic and disconnected footpaths making walking unattractive and unsafe. For this reason it has been preserved as a walking city, and is a shopper's delight. This city is not built for cars and thus cannot live easily with large numbers of them.

The chief attribute of a compact development is that it has a greater density than a suburban development and uses land to maximum efficiency. The population density varies from 150 to 450 persons per ha, depending on the vertical expansion of the buildings. Its primary attributes are high density and mixed land use so that people can live near their workplace and leisure facilities. Demand for travel is reduced and people can walk and cycle easily to their destinations. It is thus highly space efficient and little land is devoted to roads and parking, making sustainable modes of transport feasible. The sustainable use of land reduces sprawl and preserves land in the countryside. It looks to recycling land in towns for development.

In social terms, compactness and mixed use are associated with diversity, social cohesion and cultural development. Some also argue that it is socially equitable because it offers good accessibility. Compact cities are economically viable because the cost of infrastructure, such as roads and street lighting, can be shared among the larger numbers of users. The population densities are also high enough to support local services and businesses. Shahjahanabad is a compact, culturally appropriate city with interlinked open spaces serving as cultural nodes.

Transit city: New Delhi

In 1912 the architects Herbert Baker and Edwin Lutyens situated the new vice-regal palace on Raisina Hill with the new city spreading out at its feet. The processional avenue of King's Way (now Rajpath) from the Palace (Rashtrapati Bhavan) to the India Gate followed the vision of the older formal city of Shahjahanabad. But the space alongside the palace was not allotted to shops, residences and temples in the manner of Chandni Chowk. The city moved further outwards and was formally planned with a strict sense of military hierarchy. The residences for the members of the Viceroy's Council took pride of place, followed by the homes for the deputy secretaries, then the superintendents and finally the clerks. Commercial enterprises were set in Connaught Place. Huge areas were cleared of forests to provide for the bureaucracy and ruling elite, with spacious avenues and parks dominating a landscape accommodating single-storied bungalows in descending order, according to rank.

Lutyen's New Delhi was planned as a radial city with a concentric activity core devoted to civic functions. After the British regime ended, the first master plan released in 1962 developed Delhi into a bus-transit city by expanding the periphery. This was done to preserve the existing core and cater for a growing population resulting from migration and the partition of a newly independent India. People walked short distances and took buses for longer ones. Today Delhi is 1,300 km^2 and it takes more than two hours to cross it. Population density varies in different pockets. Unlike most Indian cities, traffic in Delhi is predominantly motorized. The road space is shared by at least seven different types of vehicles, each with different static and dynamic characteristics.

The proportion of fast-moving vehicles, especially light vehicles, has increased dramatically over the years. A study by the Indian Institute of Technology at 13 different locations in Delhi in 1993–4 showed that the share of non-motorized transport ranged between 8 and 66 per cent, while motorized two-wheelers accounted for between 22 and 55 per cent, and cars accounted for between 15 per cent and 44 per cent (Tewari et al. 1998). Nearly 32 per cent of all commuter trips in Delhi are walking trips. Road-based public transport, including chartered buses, accounts for 42 per cent of all trips. Of all commuter trips, around 11 per cent are by sustainable modes of transport, such as cycles and rickshaws, 5 per cent by car and 12 per cent by motorized two-wheelers. Table 11.1 shows the changing modal share of trips in Delhi with the share of trips by motorized two-wheelers increased significantly from 1981. During the same period the share of bicycle trips declined considerably.

Table 11.1 Share of transport modes in New Delhi, 1957–94

Mode	Share (%)				
	1957	1969	1981	1994	1994[a]
Cycle	36.00	28.01	17.00	6.61	4.51
Bus	22.40	39.57	59.74	62.00	42.00
Car	10.10	15.54	5.53	6.94	4.74
Scooter/motorcycle	1.00	8.42	11.07	17.59	12.30
Three-wheeled scooter taxis	7.80	3.88	0.77	2.80	1.91
Taxi	4.40	1.16	0.23	0.06	0.04
Rail	0.40	1.23	1.56	0.38	0.26
Other vehicles[b]	17.90	2.19	4.10	3.62	2.47
Walking	N/A	N/A	N/A	N/A	31.77
Total	100	100	100	100	100

Note
a Includes walking trips.
b Includes cycle rickshaws and thelas (human-powered vehicles).

Air pollution and congestion is sizeable due to inefficient bus services, bad traffic management and scant encouragement of buses, such as no policy provision for exclusive bus lanes. The government is constructing a mass rapid-transit system based on rail to reduce congestion and pollution. This is not expected to be very effective. At present, the Delhi Transport Corporation runs at least 650 bus routes in the city. A 200 km metro system could not match the catchment area covered by an extensive bus system (The Energy and Resources Institute 1994).

Studies have shown that high density must be linked to rapid transit if a city is to reduce its use of energy for transport. However, as cities also need open space, the two goals can be in conflict. Hence any transit-oriented development implies a somewhat linear urban pattern. It cannot be just a denser suburban mixed use that is located at a transit stop. Land uses in the area immediately adjacent to the transit stop tend to be limited to those that are compatible with and supportive of the transit stop, so that sufficient development intensity must be clustered immediately adjacent to the transit stop. Its vitality and success depends on having enough people using it at all hours of the day.

With respect to users, everyone who gets on or off public transit is a pedestrian, regardless of how they get to the area. Comfortable, convenient walkability to the transit stop is therefore essential. The transit stop needs to give passengers access to a convenient, integrated regional transit system that will connect them to the main destinations throughout the region. Similarly, the transit stop needs to be connected by a network of streets and pathways to adjacent neighbourhoods and allow direct access to the transit stop without relying on arterial streets. Rail-based transit is more expensive to construct than bus-based transit, but adding either increases the value of adjacent properties served by the transit (Cervero 1998; Victoria Transport Policy Institute 2007).

The automobile city: Gurgaon

Gurgaon is a satellite town of the National Capital Region of India, lying 32 km south of Delhi. Urban development in Gurgaon occurred through joint government and private-sector initiatives. This was envisaged to lead to faster and better urban growth through access to more financial resources. The demand for growth is dictated by large multinational companies, software and technology-oriented companies and upper-middle-class families who seek larger living space at affordable prices. Today, skyscrapers and modern shopping malls dot the new city, which saw a major real-estate boom in the late 1990s and since 2000.

Gurgaon has skyscrapers along its arterial roads. The local government has permitted scattered commercial and residential development in outlying areas that lack the necessary infrastructure to support it, such as roads, utilities, hospitals, shopping and schools. That generates long trips from home and to almost all destinations. Gurgaon at present offers little public transport to residents. As an automobile-oriented city it has few intercity and intra-city transport networks. Hence, the transportation network has failed and commuting is essentially by private vehicles, which increases the load on the meagre road infrastructure creating congestion, air and noise pollution and an imbalance in urban ecosystems.

Energy demand in Gurgaon is increasing at an average rate of 17 per cent p.a. with only a 5 to 7 per cent increase in supply. Added to this increasing gap between demand and supply are high transmission and distribution losses amounting to between 20 and 25 per cent, due to the large network (Jangra and Dhariwal 2002). In Gurgaon the residential and commercial sectors consume 25 per cent of all electricity, most of which is utilized in buildings in the form of air conditioning and lighting.

A conventional planned city: Chandigarh

In 1952 Le Corbusier designed a new capital for the Punjab state at Chandigarh. The site had to accommodate an initial population of 150,000 (ultimately 500,000; now one million and still growing). Le Corbusier identified the four basic functions of a city as living, working, circulation and caring for the body and spirit. To Le Corbusier, circulation was of great importance and determined the other three basic functions. By creating a hierarchy of roads he sought to make every place in the city swiftly and easily accessible and at the same time to ensure the tranquillity and safety of living spaces.

The flat, gently sloping site on which Chandigarh was built is located between two seasonal rivers some 8 km apart. The urban form of Chandigarh (tilted on a north-east/south-west axis) is a tidy chequerboard pattern adapted to the particular attributes of the site and resulting in a distinctive distribution of functions and a hierarchy of roads. The city was planned to be free of the familiar overcrowding, pavement dwellings and squatter shanties of many Indian towns. The basis for the plan was a 'sector' (subdivided into 'urban villages' of about 150 families, equivalent to an average traditional settlement found in the Punjab). A classified circulation pattern resulted from his theory of the seven Vs (*les sept voies*). A regular grid of traffic routes (V3) defines the various sectors, which are introverted self-sufficient living units connected by V4 traffic routes. The first phase of the plan included 17 sectors, each 1,200 m × 800 m in area. Shopping and bands of open space cut across each sector, fixed at some 400 m apart. Vertical green belts with pedestrian (V7) routes contain schools and sport activities. The city's V2 routes cut across these routes and consist of three major avenues: the People's Avenue, a ceremonial approach to the Capitol; the Middle Avenue, connecting the railway and the industrial area to the university; and the South Avenue, marking the boundary of the first phase of the city (Chandigarh Industrial and Tourism Development Corporation n.d.).

Chandigarh could be called a collection of villages, since land use there is highly segregated. Each sector was designed as an enclave surrounded by a green belt and with only four points of entry. As you drive through the city, therefore, you get the impression that there is ample scope for increased building density. Emptiness also prevails in the government sector. The pressure exerted on the empty space of Chandigarh by the growing agglomeration of buildings around it is increasing because of its high potential for densification. In a sense, Chandigarh can even be seen as a well-kept Central Park within the metropolis that is rapidly growing up around it. The notion that it can exist only thanks to this free zone outside the city is no exaggeration. Economically, Chandigarh is partially supported by the government by way of civil-service wages, but increasingly also by inputs from the agglomeration. So, despite its indisputable planning qualities, Chandigarh's 'success' is largely artificial (Vollaard 2003). At the city scale, the isolation of the routes and avenues, together with the zoning regulations, do not encourage intense urban activity. Roads and hard-paved plazas take up a large amount of space. The city's own rigid character, lacking urbanity, presents an image of a vast series of urban hamlets. In terms of mobility there has been provision for a grid of fast traffic roads. The roads are too wide even today, when the population has far exceeded the target of half a million (Vollaard 2003).

Even though the master plan incorporated a fair share of green areas and the traffic system also has a hierarchy of roads ranging from highways to cycle tracks, there are problems with the city planning. The pedestrian paths are not popular as they are not well-shaded by trees. The city has huge expanses of hard surface. This has brought microclimatic changes. Problems of water percolation and flooding are also associated with the impervious road surface. As a result the water-table level has dropped over the years.

CONCLUSIONS

A comparison of these habitat patterns, built in a similar cultural and ecological context, and ranging from an indigenous city and a transit city, an automobile city to a unique experiment in a planned city built on a greenfield site, leads to some useful conclusions. There is a critical need to develop a habitat model that synthesizes the attributes of the various habitats discussed above to meet the needs of a sustainable city. Current and future cutting-edge technologies and information-system technologies that are completely reforming the manner and place of work may help to redefine conventional physical boundaries and the relationships between places of living and work. Redefining the various physical elements of the habitat pattern can re-establish relationships to land, the most fundamental of these being natural resources. Food production can perhaps take a new form in urban agriculture.

The integration of contemporary modes of transport can help to achieve this symbiosis with the minimum consumption of natural resources. As a result urban patterns could once again lead to a symbiosis of nature, people, technology and human aspirations. This is a new era in which built form can be developed to respond to contemporary and future ecological parameters: a new age in architecture is waiting to be explored. Design is a very powerful tool that can achieve balance between man and nature. The only limit to what can be achieved is our own imagination.

The future city

A possible future city model is presented in Figure 11.20. This is by no means an attempt at developing a master plan. It redefines relationships between the

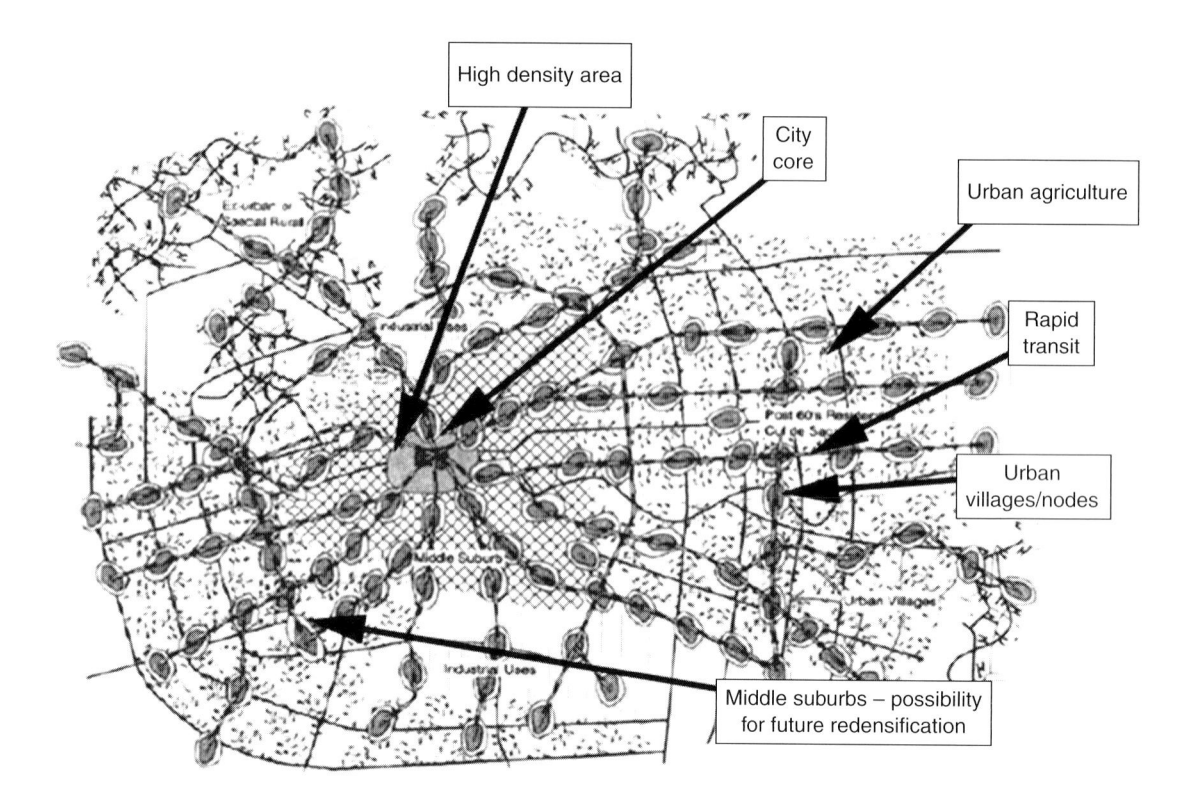

Figure 11.20 Possible future information city – urban villages/nodes (walking-oriented) linked by quality public transport.

place of residence and that of work, based on information-system technologies, redefines and re-establishes relationships between built forms and nature and minimizes environmental threats and stress caused by transportation. This model can help to minimize a city's ecological footprint and optimize its use of natural resources. It could help to generate a culturally appropriate habitat and built form.

Eco-city concepts: the Delhi context

The study areas were chosen for their inherent assets, in the form of either their natural resources or the human modification of these resources that has made them indispensable and more prone to depletion. The city of Delhi is not well-known for its eco-assets because we chose to neglect, over-use and erode its natural resources in direct opposition to the concepts of sustainability. The pollution of the Yamuna River is one of the prime examples of this tragic destruction of our natural resources. The central ridge is another example of our ecological insensitivity by understating its potential as a watershed, a bio-diverse ecosystem and the central relief to the city. Any urban model to suit the city of Delhi must address these ecological danger areas in order to realize any goal for sustainability.

Figure 11.22 Gol Market, New Delhi, India – green web.

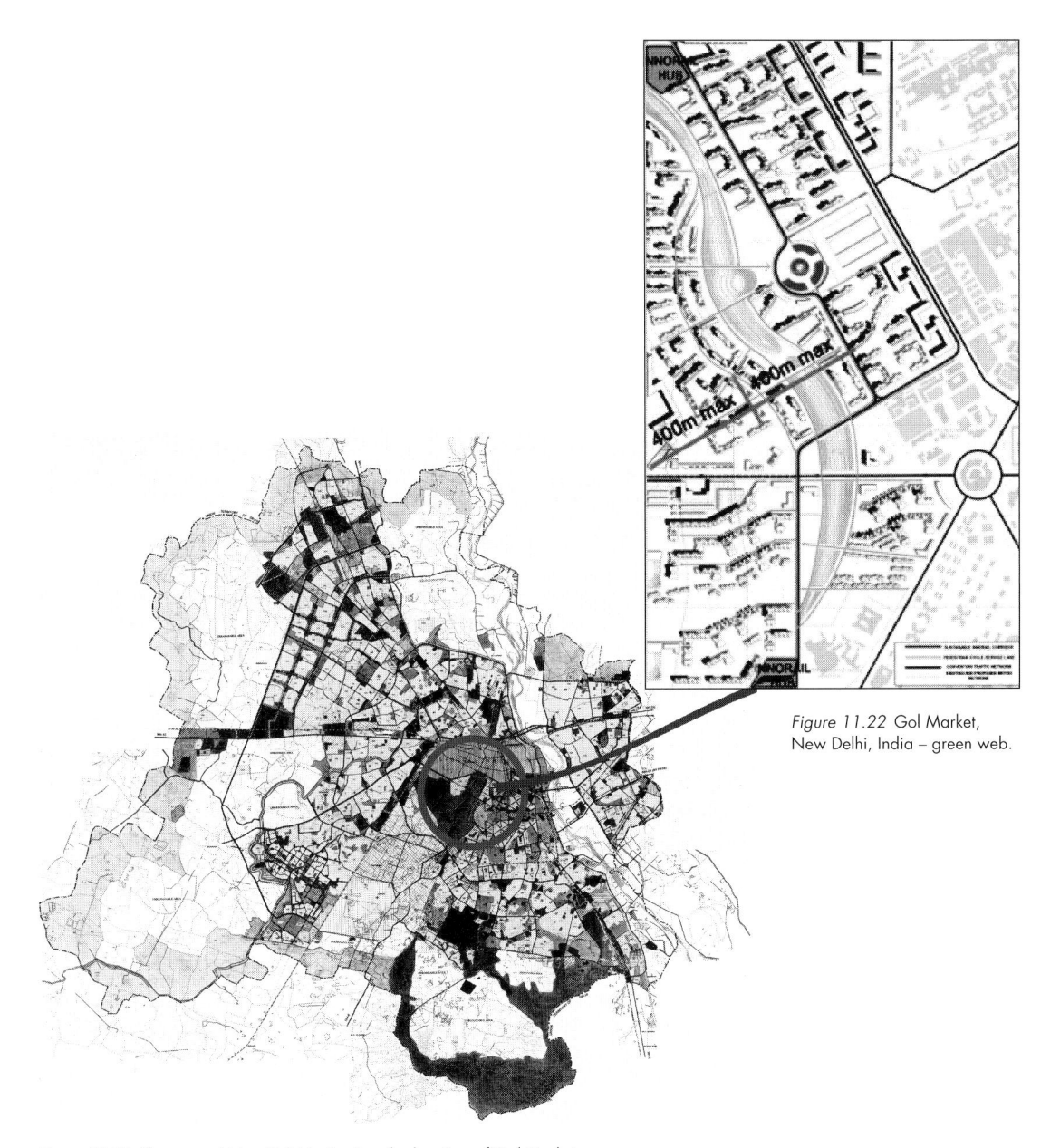

Figure 11.21 The map of New Delhi indicating the location of Gol Market.

Figure 11.23 Gol Market, New Delhi, India – 3D model (see also Figure 11.30, colour insert 2).

Figure 11.24 Rajendra Nagar, New Delhi, India – 3D model.

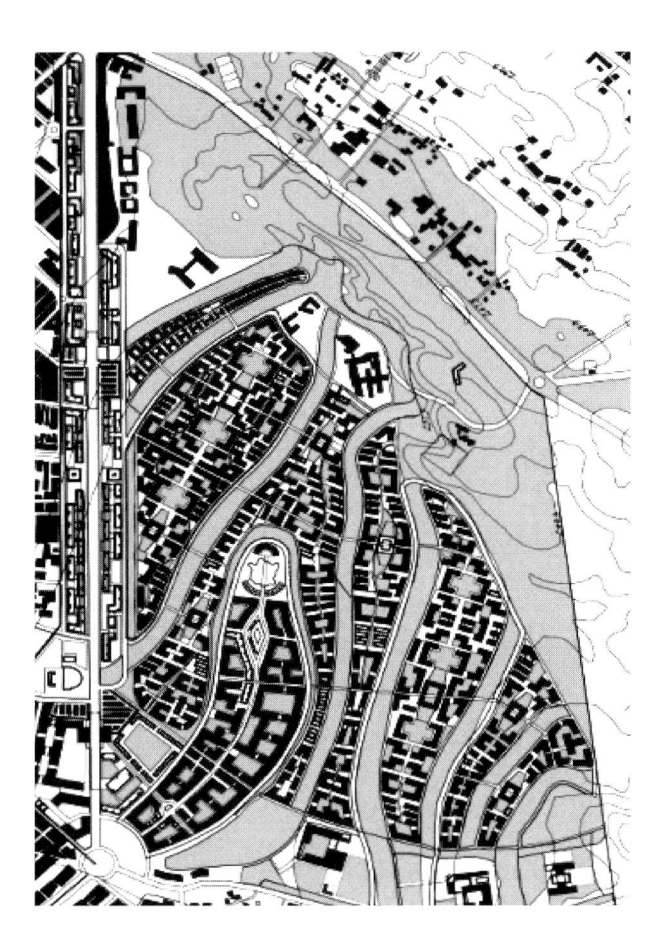

Figure 11.25 Rajendra Nagar, New Delhi, India – the proportion of built and unbuilt structure.

Figure 11.26 Rajendra Nagar, New Delhi, India – 3D view.

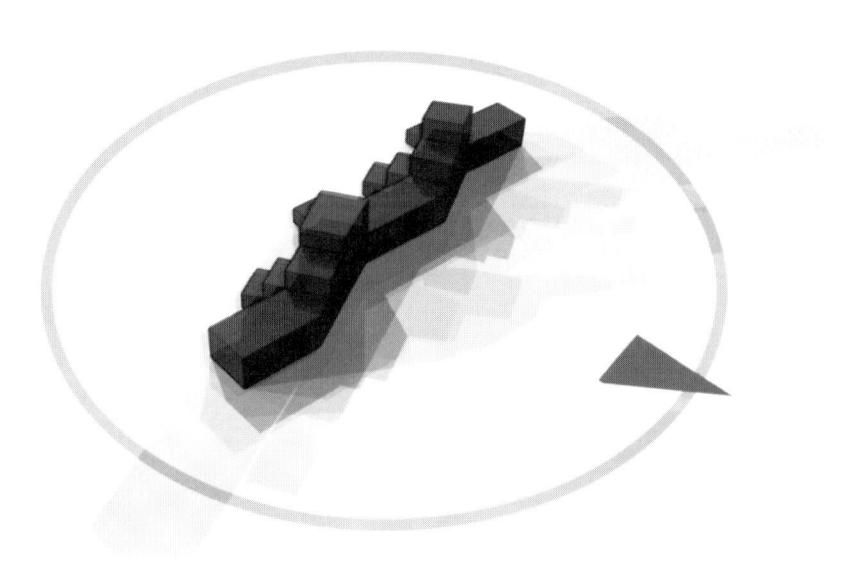

Figure 11.27 Gol Market, New Delhi, India – solar study of linked forms.

ECO-CITY? ECO-URBANITY?

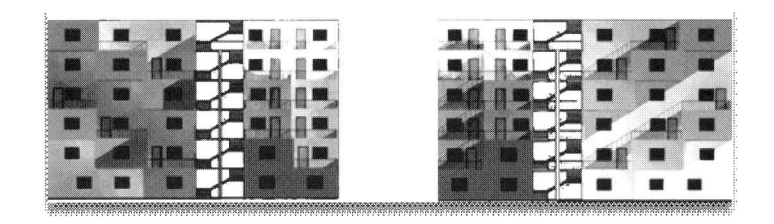

Figure 11.28 Rajendra Nagar, New Delhi, India – elevation, plan and 3D view.

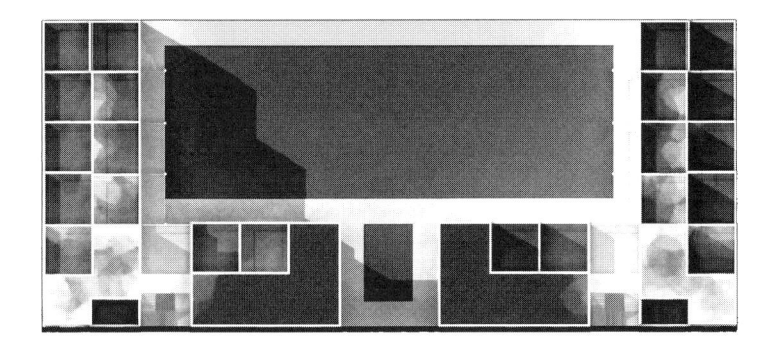

Figure 11.29 Rajendra Nagar, New Delhi, India – 3D view.

BIBLIOGRAPHY

Brundtland Commission (1987) *Our Common Future*, Oxford: Oxford University Press.

Cervero, R. (1998). *The Transit Metropolis, a Global Inquiry*, Washington, DC: Island Press.

Chandigarh Industrial and Tourism Development Corporation (n.d.) Homepage. Available online: www.citcochandigarh.com (accessed 19 May 2008).

Jangra, V.K. and Dhariwal, R.K. (2002) *Haryana Urban Development Laws*, Chandigarh: V.K. Jangra.

Johnson, C. (1969) 'Geometries of power, imperial cities of Delhi', *The Wilkinson Lectures*, pp. 31–62. University of Sydney.

Krishan, A., Tewari, P. and Jain, K. (eds) (1995) *Climatically Responsive Energy Efficient Architecture: A Design Handbook*, Vol. 1 and 2, New Delhi: Center for Advanced Studies in Architecture, School of Planning and Architecture.

Radović, D. (2007) 'Introduction', *Eco-urbanity Symposium*, 7–8 September, Toyko: cSUR Centre for Sustainable Urban Regeneration, University of Tokyo.

Tewari, G., Mohan, D. and Fazio, J. (1998) 'Conflict analysis in mixed traffic conditions', *Accident, Analysis and Prevention* 30, 2: 207–15.

The Energy and Resources Institute (1994) New Delhi: TERI.

Victoria Transport Policy Institute (2007) 'Transit oriented development, using public transit to create more accessible and livable neighborhood', *TDM Encyclopedia Online*. Available online: www.vtpi.org/tdm/tdm45.htm (accessed 19 May 2008).

Vollaard, P. (2003) 'Chandigarh in intensive care', *Archis*. Available online: www.archis.org/plain/object.php?object=871&year=&num= (accessed 19 May 2008).

Part III

OTHER SCALES AND SENSIBILITIES

INTRODUCTION

Krishan's case studies bring us to the third cluster of chapters, one that focuses on environmentally and culturally sustainable architecture. Arie Rahamimoff opens the cluster with a reminder that urbanism and architecture based on ecological principles have a long history. He focuses on the region in which he lives and in which much of his work was developed and points to the masterpieces built in Middle East by King Herod, in response to the climatic and environmental conditions of Jerusalem, the Mediterranean coast and the Israeli desert. The Nabateans who lived along the main Incense and Spice Route between Arabia, India and the Mediterranean established over 1,000 years a culture that flourished in the extremely arid conditions of the desert. They developed cities, farms and agriculture that was highly adapted to its context. During the Ottoman Empire, the cities in the region continued to be built on principles of ecological town planning, while in the last century, the Israeli version of Bauhaus in the White City of Tel Aviv is an example of the best of what modernism had to offer – sound town planning and architecture based on the specific conditions of the region. Rahamimoff demonstrates the continuity of those concepts in his own practice. His examples include residential neighbourhoods, academic campuses and public buildings. Those projects utilize timeless principles, including the treatment of the rivers that give form to urban structure, attention to the trajectory of the sun, the use of solar envelopes for compact urban forms, the collection of run-off water, vegetation as a key architectural feature for human comfort, and the economics of energy.

Kengo Kuma argues in favour of bringing back nature and history to the city in a very specific way. The opening position of his predominantly visual chapter is that an effective method of bringing nature back is to use natural materials. This simple argument is based on an irony: Tokyo, Kuma reminds the reader, used to be a city where architectural expression was constrained by limits of its key building material – wood. As with all natural materials, wood embodies specific positive restrictions that contribute to the definition of urban character. They determine the scale of the architecture and the streets and, ultimately, generate human-sized environments. Kuma invites us to see the entire architectural culture of Tokyo as born from the restrictions of the wood with which it was built. With the introduction of concrete, such restrictions disappeared and, as a result, the character of Tokyo was ruined, the human scale was lost and the humane culture of Tokyo was all but destroyed. Kuma believes that by reintroducing wood into the city, we can start harbouring hope for regenerating some aspects of Tokyo's culture. A similar logic applies to urban vibrancy, where the author proposes various reinterpretations of time-tested ways and practices of (re)igniting urbane and liveable public life. He argues that, while the main themes for urban planning in the twentieth century were expansion and functional zoning, for the twenty-first century it should be the reproduction of rich, multi-purpose spaces that bring together a variety of central urban functions.

Kodama Yuichiro brings to our attention some issues associated with the use of renewable energies in contemporary buildings. He observes that buildings increasingly use air-conditioning and reminds us that in Japan, more than one-third of global-warming emissions comes from the building sector. At another level, the lavish use of air-conditioning in design also contributes to the separation between indoor and outdoor spaces and between public and private realms. Being controllable, the indoor climate begins to be seen as safer, and the outdoors is endlessly exploitable and polluted. What Kodama calls a positive spiral structure is needed to replace the dominant, negative spiral syndrome, where energy is consumed to make the indoors comfortable while the environment suffers. A positive spiral seeks the coordinated improvement of both the indoor and outdoor environment. The use of natural daylight, solar heating, and the very feeling of a breeze in an indoor space makes it lively and refreshing and, consequently, conducive to better work conditions and efficiency. The chapter argues that creative integration of architectural design and advanced technology are required, along with flexible and innovative design methodology. Renewable energies provide a solution. They are clean and, if rigorously applied, can also generate new designs to achieve new types of low-energy architecture.

The final chapter in this cluster focuses on constructions that may be the smallest in scale, but are of immeasurable significance: graves. Naito Hiroshi's largely visual chapter addresses the topic of death and cultural belonging. He uses the examples of cemeteries in Japan to explain the relationship between the form these burial sites take and contemporary Japanese. Death brings each of us down to the very essence of our biological and cultural self. The argument of Naito's chapter is that through thinking about death we can obtain new insights into the ways we live and establish wider and fresh perspectives on cultural belonging. He uses three of his recent projects to locate death in its cultural context, and to refer to some of the current discussions about the essence of Japanese culture and Japaneseness.

12 ECO-URBANISM
An Israeli perspective

ARIE RAHAMIMOFF

INTRODUCTION

Urbanism and architecture based on ecological principles have a long history in the Middle East. King Herod (first century BC) built several masterpieces that responded to the climatic and environmental conditions of Jerusalem, the Mediterranean coast and the Israeli desert. Among the major features of his responsive architecture were an excellent choice of location and orientation, a perfect definition of indoor and outdoor spaces and careful collection of water run-off. The Nabateans lived along the main Incense and Spice Route between India, Arabia and the Mediterranean. Over a period of 1,000 years ending in about the eighth century AD they established a culture that flourished in the extremely arid conditions of the desert. Here they developed agriculture, farms and cities that were very well adapted to the desert context. Later on, the cities built under the Ottoman Empire were based on sound principles of ecological town planning. Focusing now on modern times, the Israeli version of Bauhaus in the White City of Tel-Aviv, built during the last century, is a shining example of appropriate town planning and architecture based on the specific conditions of the region.

Over the last 30 years I have practised architecture and town planning based on ecological and environmental principles in order to establish a proper relationship between the built form and the urban and regional context. This chapter presents the architectural and urban concepts that were developed as a result of my experience with an urbanism based on ecological

Figure 12.1 Compact morphology of the Old City of Jerusalem responds to the climatic conditions of the desert setting.

principles, desert architecture and the compact configuration of built form. This was expressed through residential neighbourhoods of various sizes in different locations that were formed to fit the specific climatic, environmental and ecological conditions of the region, as well as academic campuses and public buildings that were built on diverse architectural and urban scales.

The principles of planning that guide the work of our architectural group fall into three categories:

- *Eco-urban structural principles.*

    River basins and mountain tops establish the appropriate topographic structure of the built form. The solar envelopes and the trajectory of the sun are key factors in developing compact urban settings. An urban and regional run-off regime is vital for a sound, sustainable environment.

- *Eco-urban cultural principles.*

    Public, urban and regional spaces are crucial in sustaining tangible and intangible culture. Architecture is based on the defined specificity of place and the local culture. An essential quality of proper sustainable architecture is the response to climate conditions and cultural contexts.

- *Eco-urban ethical principles.*

    Shortage of land, water and energy can catalyse and stimulate architectural creativity. Architecture and urban design bear multigenerational responsibility. The combination of bottom–up and top–down planning is indispensable for the long-lasting and successful development of communities, both urban and regional.

Combining these principles in various ways creates diverse and rich architectural forms. These fundamental aspects of contemporary planning and design are the driving force behind our work. We derive our inspiration from past experience and future needs.

THE CONTEXT

The Middle East is the meeting point of Africa, Asia and Europe. Within this limited geographical area we have physical contexts of a great variety. They provide opportunity for us to explore solutions for built form in the mild coastal climate along the Mediterranean, as well as under extreme, harsh desert conditions similar to those of Jerusalem only a few kilometres away. The rich history of our region supplies us with many superb examples of responsive architecture and urbanism. Then as now, it is the blend of architecture, and urban and regional planning that gives the right answer for constructing a well-mannered built environment.

For example, the Old City of Jerusalem is a compact, dense urban form that responds perfectly to its context in the Mediterranean desert. The principles listed in the ECO-Urban Ethic category can easily be applied to the functional design of the Old City. This is especially evident in the way they solved the problem of supplying water to a population that was totally dependent on the successful design of the water system. Scarcity of water has always been and is still today an integral factor in shaping urban form, determining whether or not a community can thrive.

As architects we are approaching a moment in the history of planning when we have to reconsider our basic planning starting points and reshape the built environment in order to optimize the usage of energy. In reshaping our built environment water will become an even more critical component. However, the Middle Eastern experience over thousands of years of building compact cities provides relevant examples for future construction.

HISTORICAL EXAMPLES OF ECO-URBAN PLANNING

Each of the following ancient sites exemplifies a historical understanding and appropriate use of the eco-urban planning principles listed above, emphasizing one or the other according to the environmental, cultural and ethical needs of the time.

The Old City of Jerusalem

The Old City of Jerusalem is confined behind thick walls that were obviously built for defence. However the compact, dense urban form built within the walls responds perfectly to the extreme climatic conditions of a city that faces the desert. A whole vocabulary of built form, such as alleys, internal courtyards, thick walls, small windows and domed roofs, voices an appropriate response to this particular Mediterranean desert condition (Figure 12.1).

Scarcity of water was a major factor in shaping the built form. Only with a reliable water system was the city able to flourish and grow. Every drop of water was drained into cisterns designed to collect the winter rain for an entire year's use. Supplementing this, an aqueduct more than 60 km long brought water by means of gravity from remote springs into the city. This remained the city's water-supply system until just one generation ago. Unfortunately, successful regional and urban planning principles that prevailed for more than a millennium were neglected when modern water-supply systems were introduced. In Jerusalem's Old City the principles of eco-urbanism are expressed in the work of these ancient planners, as exemplified by its water system and existence over many generations.

Masada

A notable example of superb built form in an extreme environment is Masada, constructed on an isolated mountain top by King Herod in around 35 BC. By constructing his palace (known as the Northern Palace) at the top of the slope on the north face, the builder avoided the overheating that was natural for the east, south and west faces, maximized ventilation and benefited from the annual mean temperature of the rock into which the palace was integrated. Rainfall was collected in cisterns on the western slopes of the mountain and

Figure 12.2 Partial reconstruction model of Masada (for aerial view, see Figure 12.3, colour insert 3).

gravity was exploited to supply water to all three levels of the building. Masada was a masterpiece of planning and was fully integrated with the environment, responding to and benefiting from its unique context (Figure 12.2).

Nabatean planning

The Nabatean culture flourished at the edge of the Roman and Byzantine empires from around 300 to 700 BC. This merchant society that traded goods transported from Arabia and India to the Mediterranean along the Incense

Figure 12.4 Reconstruction of a Nabatean farm in the Israeli desert.

and Spice Route is one of the most complex desert cultures known (Figure 12.4). The Nabateans developed unique regional, urban and architectural methods of construction that perfectly suited the extreme conditions of the desert. With a rainfall of less than 50 mm per year and the annual temperature fluctuating from almost freezing to 40°C, the Nabateans had to shape their regions, cities and houses with great sophistication.

In order to sustain agricultural production, they collected run-off water using highly complex methods. All of the *wadis* (dry riverbeds) were terraced. Stones were gathered from the slopes and used to direct run-off water to the terraces, multiplying in this way the amount of water that reached the terraced *wadis* 30-fold. As a result, in this arid zone rich orchards and plantations in this arid zone were able to supply food for a large population.

The Nabateans also demonstrated careful integration and a use of resources that was extremely well-adapted to the terrain. Their temples and public institutions were hollowed out of the rock, and the stone thus mined from the mountain was used to supply building material for residential neighbourhoods (see Figure 12.5, colour insert 3). In fact, the amphitheatre was also a quarry.

Individual buildings were built with thick walls that, in the summer, absorbed the diurnal heat and later, during the cool of the night, radiated that warmth into the house. So that over a 24-hour day throughout the entire year the time lag of thermal mass provided human comfort. The extremely small windows reduced glare and heat gain in the summer and decreased heat loss in the winter. In sum, the Nabateans expertly implemented the very principles that modern architectural planning must adopt in order to facilitate harmony between total urban existence and arid environments.

Tel Sheba

The ancient city of Tel Sheba is another fine example of sustainable city building in arid zones (a *tel* is an artificial hill formed of cities built one on top of the other following serial destruction and reconstruction).

Throughout the various periods of civilization evident in this *tel* located on the edge of the seasonal Hebron River, a unique system of collecting water was employed. When there were floods the riverbed was dammed and the water diverted into a cistern. From that cistern the city's population was supplied with water throughout the year. Two additional water sources completed a three-tier water-supply system for this ancient arid site. In addition to diverting the flood water from the river, run-off water was collected and ground water was drawn from wells. This composite method of controlling water resources was their key to survival. In 2003, Tel Sheba was declared a World Heritage Site by UNESCO (Figure 12.6).

URBAN PLANNING

At the beginning of the twentieth century the White City of Tel Aviv rose from the sands of the Mediterranean shore. Here the Israeli Bauhaus archi-

Figure 12.6 Tel Sheba: a Tel is a multilayered ancient city. Tel Sheba conceals an underground water reservoir.

tects responded to the spirit of the socialistic era of the new society in their architecture and town planning. Apartment blocks were built so that each apartment opened onto spacious terraces that enjoyed the sea breeze. Large windows enhanced ventilation and cantilevers supplied shade. Carefully selected trees cast their shade, improving the micro-climate of the outdoor spaces. More than 1,000 buildings of this Mediterranean Bauhaus type form the World Heritage Site of the White City of Tel Aviv.

MODERN PROJECTS BASED ON ECO-URBAN PLANNING PRINCIPLES

As professionals we have accumulated some experience in the application of eco-urban principles in the arid climate of Israel. A selection of projects, completed and in progress, is described below, focusing on the unique challenges of each.

River planning. Rivers are the lifeline of many cities. In Israel the riverbeds were a neglected resource until ten years ago, attracting the most negative elements of modern environment and society. Sewage and garbage had accumulated in the unkempt surroundings along the riverbeds. This continuing deterioration has been reversed over the last ten years by implementing an integrated river planning project. The major objective was to rehabilitate the environment within and along the rivers and to return these linear urban spaces to society and to nature. The two different urban contexts addressed by our team are described below.

The Kidron Valley

The Kidron Valley is located between the Old City of Jerusalem and the Mount of Olives within the borders of the city of Jerusalem. This cultural space, sacred to three monotheistic religions, was practically inaccessible to residents, tourists and pilgrims. A continuous planning process was then initiated and supported by the government, public agencies and donors, resulting in the construction of promenades, observation points and trails. This unique site has now become accessible to all and has become a popular focal point for urban activity (Figure 12.7, colour insert 3).

The Yarkon River

The Yarkon River provides the most important metropolitan open urban space for central Israel. About two million people live along this river. However, for two generations this space was the neglected backyard of metropolitan Tel Aviv and was characterized by poor environmental and social conditions. Through the implementation of our plan, this situation has been reversed. Public urban institutions are now attracted to the banks of the river and 'green fingers' connect the inland areas to the river. A total of 28 km of riverbanks connects the mountains to the sea and forms a link between seven municipalities. The rehabilitation of the river has marked the renaissance of the area's urban life (Figure 12.8, colour insert 3).

Beersheba Old City

Over a period of about 6,000 years the city of Beersheba (also called the City of Abraham) developed along a dry riverbed. This shallow riverbed connects different periods of the development of the city and yet today is being transformed into the main urban and regional axis of modern metropolitan Beersheba. The archaeological traces of the Old City of Beersheba (Figure 12.9) date back to the chalcolithic period (4000 BC), since when there has been continuous settlement along the banks of the riverbed. The recent orthogonal grid of the city is from the Ottoman period of the early twentieth century and is a perfect example of a compact design for urban living. The urban grid is composed of 100 urban blocks that are divided into 16 plots, all slanted towards the north at an angle of 45°. Beersheba is an interesting example of compact city building that expands the concept of sustainability into modern times (Figures 12.10, 12.11 and 12.12, colour insert 3).

We are currently involved in preparing the master plan and detailed plans for the Old City. The purpose of the master plan for the city is to allow modern living in the historic urban context. While preserving the urban structure with stringent conservation guidelines, it provides for the introduction of moderate traffic control, separating urban and local traffic, and prioritizing

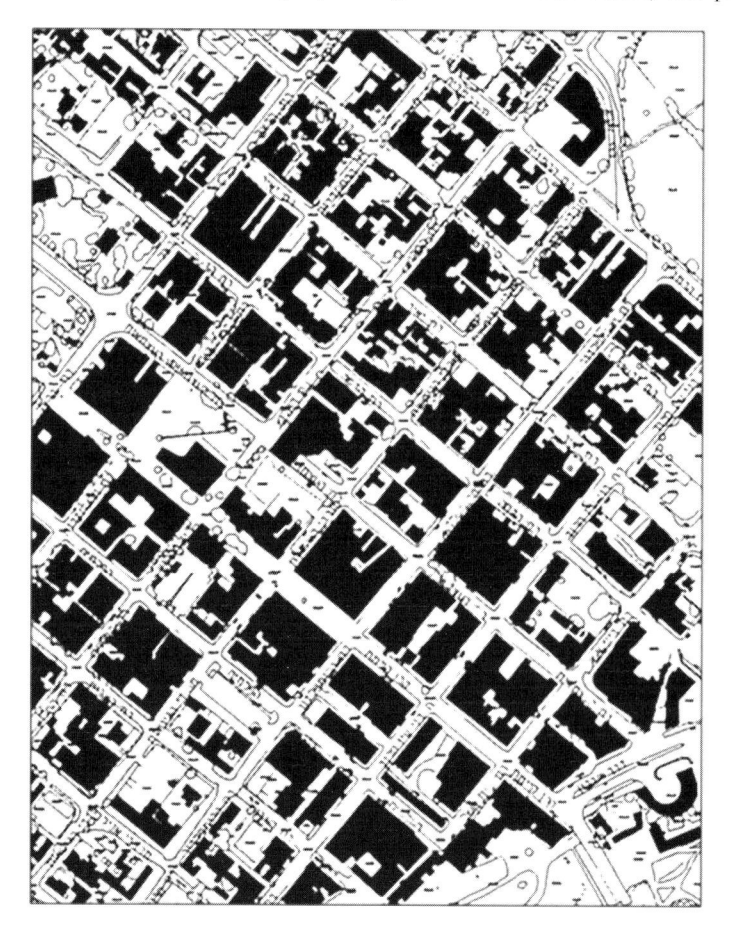

Figure 12.9 Orthogonal grid of the Old City of Beersheba.

pedestrian and bicycle movement. The plan imposes solar rights within a compact urban morphology.

The plan proposes to double existing building rights in the Old City while maintaining its character and structural morphology. It recommends the use of highly insulated building solutions for walls and roofs and carefully defines the optimal size of windows. This will increase winter radiation into the building while minimizing summer overheating. The concept of internal patios, roof gardens and terraces within a compact city structure forms a viable planning principle for many urban environments in arid and semi-arid zones around the world and is similarly proposed in the master plan for the Old City of Beersheba.

A new urban district of 10,000 dwelling units was planned and has been under construction over a period of 15 years in the northern part of Beersheba. One important strategy in the master plan is the conservation of run-off water and its redirection for irrigation. The Ramot Rabati plan has implemented some of the lessons learned from studying the design of the Old City of Beersheba. The urban blocks define the street façades. The blocks are angled to the north to increase ventilation and to allow solar radiation to permeate the length of all streets in the winter. The principle of solar envelopes was systematically adapted by constructing the lower buildings on the southern part of the block that rises towards the north. The drainage pattern of the entire neighbourhood is directed towards the open spaces in the valleys in order to minimize waste of run-off water and to reduce the amount of artificial irrigation needed for vegetation in the parks. Such an urban morphology assures human comfort and reduces dependence on non-renewable energy resources both in winter and in summer.

Albert Katz International School for Desert Studies and Neve Zin

Two projects were built during the last 15 years in the Israeli desert; a community of private homes for scientists and the International School for Desert Studies branch of the Institute of Desert Research in Sde Boker. These medium- and low-density communities follow the same standards as the projects described, but on a smaller scale. The scientific community of Neve Zin was planned in line with the concept of architectural diversity. In spite of the variations between the buildings, all were constructed following the same basic planning principles, in accordance with eco-urbanism principles for arid zones (Figures 12.13 and 12.14, colour insert 3).

The largest house in this neighbourhood was designed for a solar scientist and a theatre director and includes a private theatre. The smallest house is only 60 m^2 and was designed for a leading Israeli environmental sculptor. The houses are all oriented to face the south-east in order to maximize their absorption of winter solar radiation. Small northern courtyards supply a shady environment in the summer while solar envelopes ensure that no building casts shade on the one behind it. Well-thought-out planning of the type and location of vegetation increases human comfort in winter and summer. Deciduous trees cast shade in the summer while allowing solar radiation to penetrate in the winter. The vegetation improves the overall outdoor climate, improves air quality by filtering dust and reduces the glare of the bright sun (Figure 12.15, colour insert 3).

The Albert Katz International School for Desert Studies accommodates academic functions and residences for 120 young scientists from more than 30 countries. This medium-density community is organized in a concentric way around the public areas of the institute, with buildings spaced to ensure equal distribution of solar radiation. The public building has in its centre a cooling tower that vents the hot air and introduces natural light into the core of the building. All these planning principles, combined with intensive insulation in walls and roofs, thermal mass and precise dimensioning of south-facing windows, create an almost zero-energy building in the extreme summer and winter climatic conditions of the desert.

Centre for Jewish Studies

The venue for the largest concentration of scholars of Judaism, the Centre for Jewish Studies was built to facilitate their work. This compact building includes 20 research units, libraries and auditoriums, all surrounding a central tiered hall. The central space is the centre of informal academic activity and supplies natural light and ventilation. Since it faces south, it allows controlled solar radiation into the building during the cold months of the year. The area is 18 m high and enhances natural circulation, thus minimizing dependence on air conditioning. The building itself is terraced so that the nine-storey building blends easily into the topography of the Jerusalem hills (Figures 12.16 and 12.17, colour insert 3).

Eilat

The port city of Eilat is located along the coast of the Red Sea in the hottest and driest region of Israel, with maximum average temperatures during summer days of about 35°C and rainfall of about 30 mm a year. We were commissioned to prepare a master plan for a community of 5,000 families appropriate for these extreme climatic and environmental conditions. In line with eco-urban planning principles, our plan was formed around two central guidelines. First, the dry river which transects the neighbourhood is the spine of educational and cultural activities and tourist attractions, and is the major public open space. Second, it was decided that the formation of a compact dwelling environment was essential for these harsh climatic conditions.

We applied a variety of planning strategies in order to maximize human comfort and at the same time reduce dependency on non-renewable energy sources such as air-conditioning. Double layers of shade reduce the extreme solar radiation. To create this double layer, a canopy of palm trees creates the first sun filter and below it grow deciduous trees, pergolas and creepers for an immediate layer of protection from the sun. In order to maximize outdoor and indoor comfort and ensure an appropriate environmental response to climatic conditions, we developed a morphology of compact housing of differing densities (Figure 12.18).

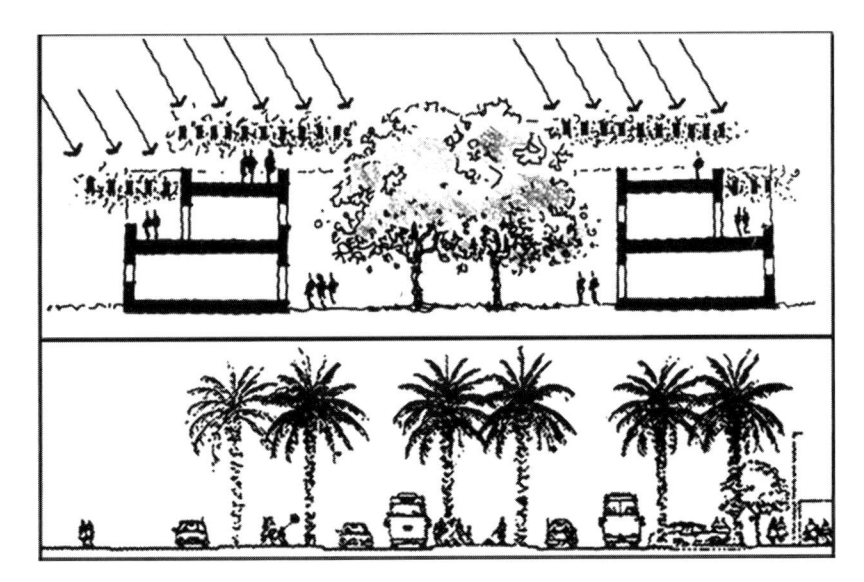

Figure 12.18 Double solar protection on urban and architectural scales is crucial for urban comfort in extreme hot and dry desert morphologies of compact desert housing in Eilat.

Though rainfall is scarce, when it rains it floods. In some cases the entire annual precipitation may occur in one downpour, resulting in one large flood for the whole year. Accordingly, we planned the geometry of the urban form to restrain the energy of water movement and to reduce land erosion. Thus, unlike the more humid parts of the arid Israeli desert where the proper urban response is to collect the run-off water, in extremely arid zones with a high probability of flooding, neighbourhoods are actually designed to act like sponges in order to absorb the majority of the rainfall. In this way, most flood damage can be avoided.

Nazareth: the transformation of a quarry into an urban block

Nazareth, the place where Jesus spent most of his life, is sacred to Christians. At present it is the largest Arab city in Israel with 70,000 inhabitants. South of the city an abandoned quarry of 30 ha forms a huge hole in this sensitive landscape. This dilapidated quarry is one of the biggest ecological and environmental hazards in the north of Israel. The negative effect of this major urban and environmental blight is magnified because it is sandwiched between Arab Nazareth and Jewish Nazareth Ilit. Worse, the new main entrance to the city runs right through the quarry.

Our plan deals with some of the major ethical, cultural and environmental issues of eco-urbanity. We proposed converting this urban and environmental hazard into a compact urban residential neighbourhood, employment area and centre of high education that will reshape the new entrance to the two cities. Buildings on each side of the main entrance are terraced down to the education centre located on both sides of main road. This plan is our attempt to reverse the negative impact of the quarry by transforming the wasted land into a new compact urban sector at the entrance to these two cities. The plan has been approved by all authorities

Figure 12.19 View of the topography of the Nazareth quarry.

Figure 12.20 Model of proposed eco-urban community of Nazareth showing restoration of wasted land.

Figure 12.21 Compact urban community at the new entrance of Nazareth and Nazareth Ilit.

and will, we hope, be implemented within the next few years (Figures 12.19, 12.20 and 12.21).

SUMMARY

In summary, 40 per cent of the global land mass is desert. Most of the urban settings in Israel require compact housing to respond to the desert conditions. Undoubtedly, water and energy will be key generators of urban and architectural form. Based on 30 years of experience of planning in the context of extreme environmental and climatic urban settings, I offer the following observations:

1. History and the experience of ancient societies is an enormous source of inspiration that should not be neglected in our future studies and projects.
2. It is essential for any urban planning to aim at reducing wasted land. Built environments should be compact. This is equally true for high, medium and low densities.
3. Planning on all scales should be responsive to environmental and climatic conditions. Basic architectural and planning principles require that not only individual buildings be designed to maximize human comfort and reduce energy consumption, but that entire neighbourhoods and cities should do the same. Furthermore, the development of whole regions should be based on comprehensive water- and energy-saving principles and on reduction of environmental hazards.

As shown in this chapter, eco-urban planning principles can provide the basis for encouraging entire countries to modify their national development strategies towards a higher environmental consciousness, a change that will enhance the quality of life worldwide.

13 BRINGING BACK NATURE AND RE-INVIGORATING THE CITY CENTRE

KENGO KUMA

This chapter examines two themes important to the concept of eco-urbanity: how to bring nature (back) to city centres and how to re-invigorate urban life, both by architectural means. Those two themes are illustrated by several Kengo Kuma and Associates projects.

We believe that an effective method for bringing back nature into the city is to use natural materials. All natural materials introduce certain restrictions but these restrictions are often their strengths. Tokyo used to be a city where buildings were limited by the natural constraints of the predominant building material: wood. For instance, it was always difficult to obtain pieces of wood larger than 90 mm × 90 mm × 3,600 mm. Such restrictions helped to define the specific characteristics of urban space, and the scale of the streets and architecture. Those limitations gave a human scale to old Tokyo and, before that, to the city of Edo.

Even further, one could argue there is a direct relationship between the materials used for construction and the culture of particular locations. The entire culture of Tokyo was significantly affected by the scarcity of its predominant structural material. With the introduction of concrete, those restrictions disappeared and, as a result, the personality of Tokyo was ruined, its human scale lost and its culture all but destroyed. By reintroducing wood into the city I believe we can help to regenerate the culture of Tokyo.

We made a number of efforts to test that design hypothesis. For instance, we used wood for the façade of One Omotesando Building (2003) (Figures 13.1a–e, colour insert 3). Japanese building regulations explicitly prohibit the use of wood for façades of that scale. However, by installing special drenchers (external sprinklers) we were able to fulfil the rigorous legal requirements and subsequently received special approval to use wood in this unique and sensitive urban location. Thus, One Omotesando shows that it is still possible to bring nature back to the city by combining new technology with traditional materials.

In another large redevelopment project in central Tokyo, the Midtown Suntory Museum of Art (2007), our main materials for the interior were wood (*paulownia*) and traditional paper (Figures 13.2a–h). The reintroduction of those vulnerable, traditional, natural materials was facilitated by carefully combining them with the latest technological achievements.

Besides our careful use of natural materials and cutting-edge technology in the Suntory Museum of Art, we constructed a traditional *muso-koshi*, a light-adjusting screen, using vertical ceramic louvres over an entire wall. *Muso-koshi* was a common detail in traditional Japanese private houses. A simple method was used to manipulate the ratio between the open and the closed wall surface, allowing users to respond to seasonal variations and particular weather conditions. This traditional environmental technology in the Suntory Museum creates sympathetic human spaces in a city that is dominated by the inhuman uniform spaces that have became common in most large-scale redevelopment schemes (Figure 13.2c). Furthermore, sliding floor plans create a spacious terrace with wooden decks used as an outdoor dining space in the restaurant and café. Such semi-outdoor spaces used to play an important role in both Western and Japanese cities. Recent redevelopments tend to partition space rigidly into either indoors or outdoors, disregarding the possibility of creating ambiguous semi-outdoor conditions. The Suntory garden terraces show our efforts to bring back the experiential richness of the once-common *engawa*, traditional Japanese semi-outdoor space.

If our work in Tokyo midtown was an attempt to reproduce nature in the city centre, our proposal for the town of Yusuhara in Kochi Prefecture was meant to unite architecture with nature. Yusuhara is a typical depopulated Japanese region. In the Yusuhara Town Hall (2006) (Figures 13.3a–g) the structure, the outer walls and the interior were all made from local Japanese cedar. At the centre of this building is a semi-outdoor space (Figure 13.3g). Large hangar doors mediate between the interior and the exterior to create this ambiguous space. During the spring and in the autumn the doors are kept open. During summer and in winter the service counters for the public office, the Agricultural Cooperative Society, the bank and the café are in the warm, enclosed indoor environment.

We believe that architectural methods can be used to bring life back to the city centre. Instead of producing the usual austere, functional scheme we proposed a range of new interlocking programmes. By combining several programmes it was possible to recreate rich and compact communication spaces, similar to Mediaeval Western squares or the alleys (*roji*) *of* Japanese towns. These projects show that using natural materials in architecture adds more to a design than a structural tool or a surface embellishment. The materials that surround us define our interaction with the environment. Natural materials can be used in a way which directly addresses the culture of the place, which nurtures certain values and attitudes towards the environment. In the same way, architectural detailing and styling decisions are not only technical or management procedures. If they are carefully chosen and executed they can bring about positive interactions within communities and between people and their tradition and environment. Thus, if the main characteristic of urban planning in the twentieth century was the expansion to the suburbs and the application of functional zoning, in the twenty-first century the challenge is to regenerate rich, multi-purpose central urban spaces that support a variety of functions in a human-friendly way.

14 SUSTAINABLE DESIGN TOWARDS A POSITIVE SPIRAL

KODAMA YUICHIRO

An essential condition of a future sustainable society is to have a low environmental impact. Reducing gas emissions in our daily lives is urgently necessary, especially reducing our daily energy consumption in the home and at work. Developed countries have to reduce 80 per cent of their present global-warming gas emissions by 2050 in order to maintain a sustainable Earth. Despite the variety of approaches to the concept of eco-urbanity, we all share a primary mutual concern: global warming. How can we create communities, towns or cities with a low environmental impact? Many ideas for a sustainable city, compact city, or eco-city have been presented and discussed on different scales, but most of them are in thrall to the modern energy myth, namely, that our problems can be solved by utilizing energy.

The key target that we must all aim for is energy efficiency; to minimize energy consumption without compromising performance is a goal of technology. Nevertheless, on occasion technological advances make the problem more complicated. A prime example is the relation between indoor environment (climate) and outdoor environment. As air-conditioning, a powerful system for controlling indoor climate, becomes ubiquitous, building design changes to accommodate and rely on air-conditioning. This is accompanied by an enormous increase in energy consumption in buildings so that more than one-third of the global-warming gas emissions in Japan are thought to come from the building sector. Consequently, energy conservation in buildings, including air-conditioning, needs to be urgently addressed in order to achieve a sustainable society.

While air-conditioning systems are considered to be a critical technological innovation by making the indoor climate comfortable, it puts a strain on the environment. The more an interior is air-conditioned for comfort, the more the exterior environment is damaged. Can we change this negative spiral into a positive spiral towards a sustainable society? How can we develop a sustainable design to achieve a positive spiral? Buildings are man-made environments built in nature. The relation between the indoor environment and nature out of doors has deteriorated because we have consumed much energy in creating that comfortable and safe indoor climate in the last 100 years. The safer and more comfortable the indoor climate becomes, the more polluted the outdoor environment becomes. This polarization of indoor and outdoor space sped up as air-conditioning technology became prevalent. The pressure to save energy is accelerating this tendency and helping to destroy a responsive, mutually accommodating relation between humans and nature.

Building design must be turned around to preserve a sustainable society and to recover the responsive relationship between humans and nature. We need to use natural energy resources such as the sun and wind, which are well-known, clean and renewable sources of energy and which can replace our use of conventional fossil-energy resources.

THE ENVIRONMENTAL IMPACT OF BUILDINGS

Per capita, CO_2 emissions in Japan in 1995 were 2.72 tons[2] per year, approximately one-third of which emitted from the building sector. The results of a life-cycle analysis of buildings show that 8 per cent of all CO_2 emissions are related to the construction process, including CO_2 emissions from manufacturing and transporting building materials, as well as emissions at construction sites. Some 23 per cent of total CO_2 emissions are related to the operation of buildings such as air-conditioning, lighting and heating water. A few more emissions are related to maintaining and demolishing buildings. Approximately two-thirds of the damaging environmental impact of a building comes from the building operation; therefore, saving energy during its operation is critical in reducing the environmental impact of buildings.

The factors that determine the amount of energy consumption in building operation fall into two groups. The first consists of factors related to the building itself, such its shape and layout and the materials used and structure of the building envelope. The second consists of factors related to building equipment and energy-management systems. This classification can be used to implement energy-saving strategies in buildings. One way of reducing the detrimental environmental impact of buildings is to utilize renewable energy resources instead of conventional fossil resources. The highlights of this strategy are aiming for a low-energy building design, highly efficient equipment and management systems and the use of renewable energy. These aspects are inter-related, and their integration is vital for success (Figure 14.1).

EVOLUTION OF BUILDING

The relation between the energy consumption and the evolution of buildings is shown in Figure 14.2. As evolution proceeds, energy consumption decreases. The first step is to make the equipment (the building service)

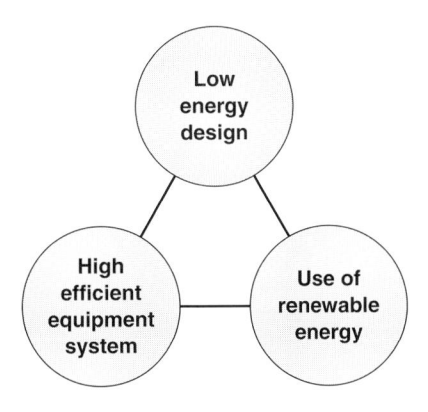

Figure 14.1 Integration of design, technology and renewable energy.

Figure 14.2 Energy consumption and evolution of building.

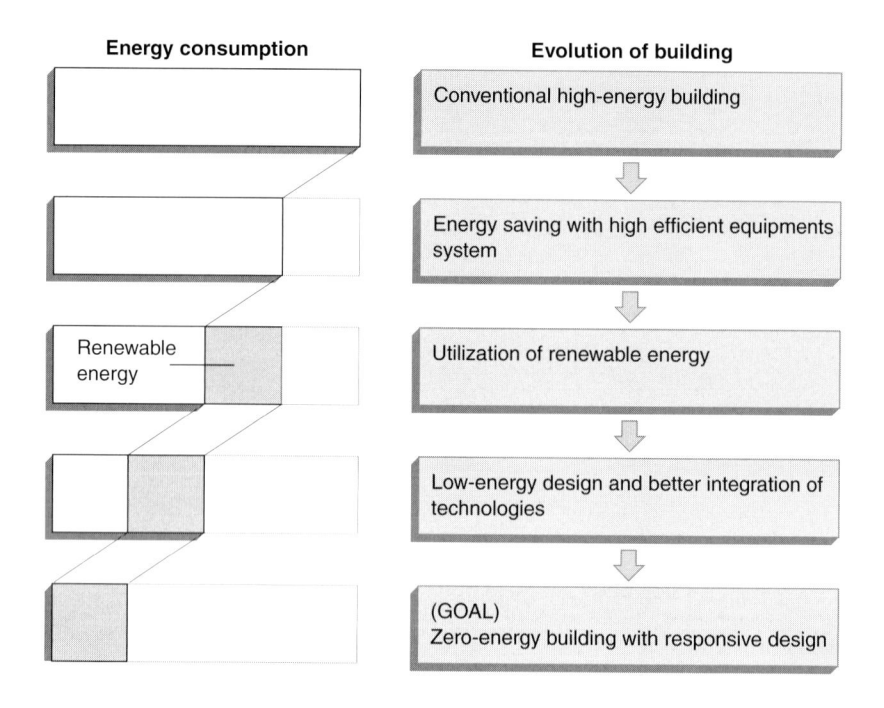

system more efficient. The second step is to replace conventional energy resources with renewable ones. In the third step, the building itself becomes the target of design and a low-energy design is considered. Other commonly used terms for low-energy design include passive and bioclimatic design. Such design shares the features of vernacular buildings that do not depend on energy to control the indoor climate. Paradoxically, today's low-energy design is based on highly advanced scientific knowledge and technologies and is supported by sophisticated design tools based on various computer-simulation techniques. Advanced low-energy design should collaborate with new advanced building-service technologies and advanced utilization systems of renewable energy. This step can therefore be a turning point towards responsive buildings in the future.

The role of low-energy design is revolutionary. It not only serves to minimize energy consumption with a highly efficient heating, ventilation and air-conditioning (HVAC) system, it also opens up the building to the outside, allowing the building once again to 'breathe' and become responsive to the outside.

POSITIVE SPIRAL

As mentioned above, the use of renewable energy is one critical concept for future buildings. However, if all that happens is a simple replacement of one energy source used in air-conditioning with another, the unfortunate struggle between humans and nature will not change. Moreover, our obsession with the effective use of energy and with which energy source is the most effective may paradoxically cause more discord between humans and nature. Low-energy design principles may help to spur a potential breakthrough and to develop the concept of eco-urbanity as well as to build a sustainable society in general.

Success may depend on constructing a positive spiral structure instead of a negative one. A positive spiral will improve both indoor and outdoor environments while maintaining responsive relations between both environments. It will also provide a higher quality of comfort to those using the building. Natural day-lighting, natural solar heating and introducing cool breezes by natural ventilation systems contribute to making indoor space more liveable and refreshing, and thus conducive to the increased efficiency of the people working in these buildings. A built example of a positive spiral structure can be seen at the NEXT21 project in Osaka, Japan. This is an experimental housing project using five ideas for sustainability:

- a long-life structural system separating the infill and cladding system from the structure;
- infill systems allowing flexible floor planning in individual houses;
- advanced energy systems for HVAC and domestic hot-water systems;
- innovated water-processing and house waste-processing systems;
- greening the building surface.

The building containing 18 units was built in 1993 and various experiments have been conducted following its completion. In 2005 low-energy design renovation was carried out as part of the third phase of the NEXT21 experiment (see Figures 14.3 and 14.4, colour insert 3).

The positive spiral was created by greening the building three-dimensionally in addition to creating a garden in the courtyard. The growth of this three-dimensional greenery generated cool spots around the building in the summer, even though the building is located in downtown Osaka where summer conditions are hot and humid. The micro-climate of the surroundings gradually changed, improving and enhancing the environment. Subsequently the inhabitants began to open their windows for cross-ventilation, especially for the cooling breeze in the evening, even in midtown where people automatically depend on air-conditioning. There was a corresponding reduction of air-conditioning heat exhaust and overall reduced energy consumption as air-conditioning was used less. The cooling effects of greening and the reduction of exhaust heat from air-conditioning are expected to create a synergistic effect. Finally, both the indoor and outdoor environment will improve and enhance the positive spiral.

15 CREATING A CEMETERY
Architecture that sustains cultural forms

NAITO HIROSHI

Cities can be seen as a metaphorical human body composed of arterial, venous and nervous systems. Enormous energy inputs (electricity, gas, water), mass distribution networks (of food and materials) and huge capital inputs (of both real and virtual money) are unceasingly pumped into the city to make the body work. The supplies constitute the urban arterial system. The energy and materials supplied are processed and stored until they need to be used and what is exhausted is recycled through the system in a new form. A kind of venous system caters for that. That is what I am interested in, especially – the role of death in Japanese culture.

Figure 15.1a The typical cityscape in Japan.

Figure 15.1b The typical cemetery in the landscape in Japan.

Birth and death are the two events in human life that are the most enriched with symbolism and meaning. To me, death marks a new beginning, in a spiritual sense: the paradoxical moment when a new arterial system commences and at the same time the physical system ends. That is when the nervous system, which aspires to control everything, finds its journey over. For me, the cemeteries are places where material body and human spirit are retrieved. If we look back at urban burial practices of the twentieth century, the age of humanism, we find that this important event – death – was socially ignored, hidden and often completely erased from sight. Through my design practice I try to bring questions of death and burial to the discussion table. In a country with an extremely low birth rate and an increasingly ageing population we need to discuss death openly so that we can better understand the ways in which we live.

Much of my recent research and design focuses on cemeteries. Parallel investigations, design and construction projects help develop a comprehensive understanding of the ways in which we deal with death and, thus, with life, by balancing the arterial and venous systems on which our existence relies. The client for the three projects that I present in this visual chapter owns a cemetery in suburban Tokyo. In the past these cemeteries were no different from those we find all over Japan. In normal cemeteries the grave plot is about 2×2 m and the name inscribed on the grave is that of the individual's family, not their own personal name. Both life and death in Japan are still strongly influenced by the traditional value that elevates family names above those of individual family members (see Figures 15.2a–c, colour insert 3).

However, that practice has begun to change. One new trend favour cemeteries without individual graves while another demands personal, rather than family, graves. Haruyo Inoue, a sociologist and the owner of the cemetery, decided to challenge these trends and create a new type of the burial

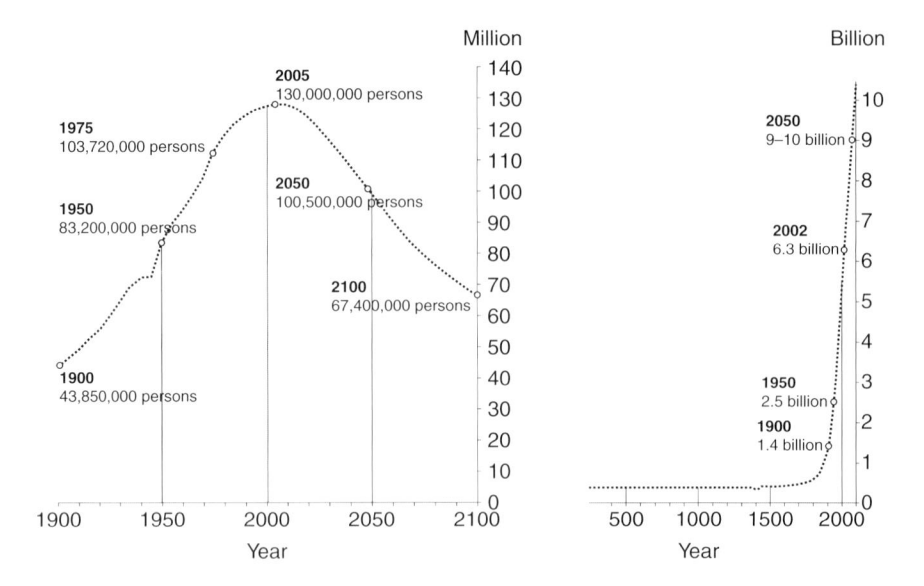

Figure 15.3 Left: The population in Japan in 200 years.
Right: The world population in 2000 years.

Figure 15.4a The first cemetery – project layout (see also Figure 15.4b, colour insert 3).

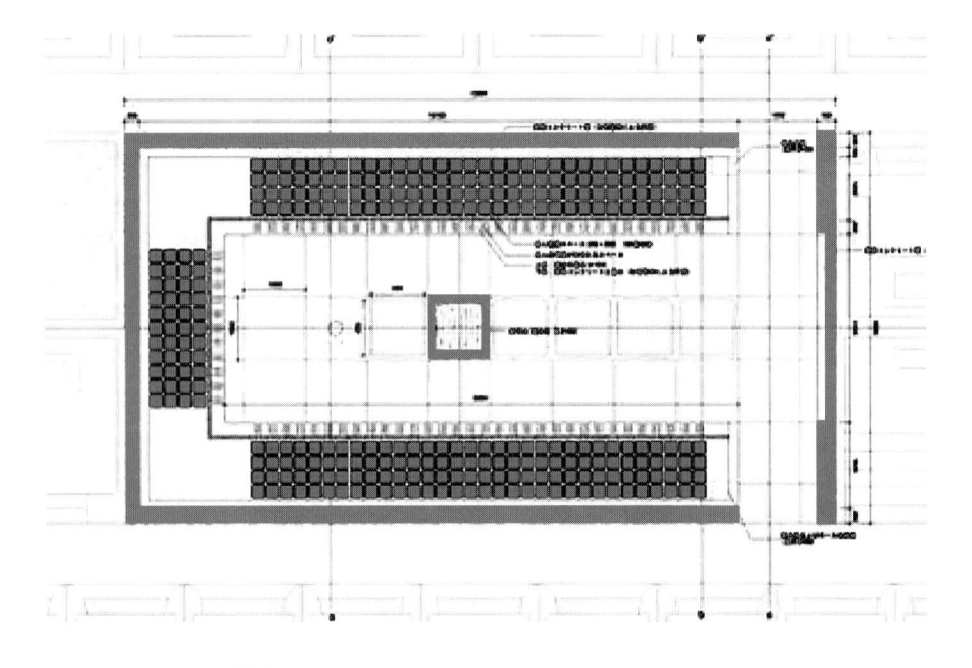

space. The concept is minimalist. The cemetery offers some cherry trees in a grass field. The burial ground has become a garden where plaques are so small that there is scarcely place for names on them to commemorate the buried person. It was this concept that offered my laboratory a design challenge.

The first of the three cemeteries occupies a narrow space between traditional graves in an established graveyard. The challenge was how to introduce the totally new burial concept and style into a distinctly conservative traditional context. We introduced walls at the edge of the plot and planted a single tree in the centre of the composition.

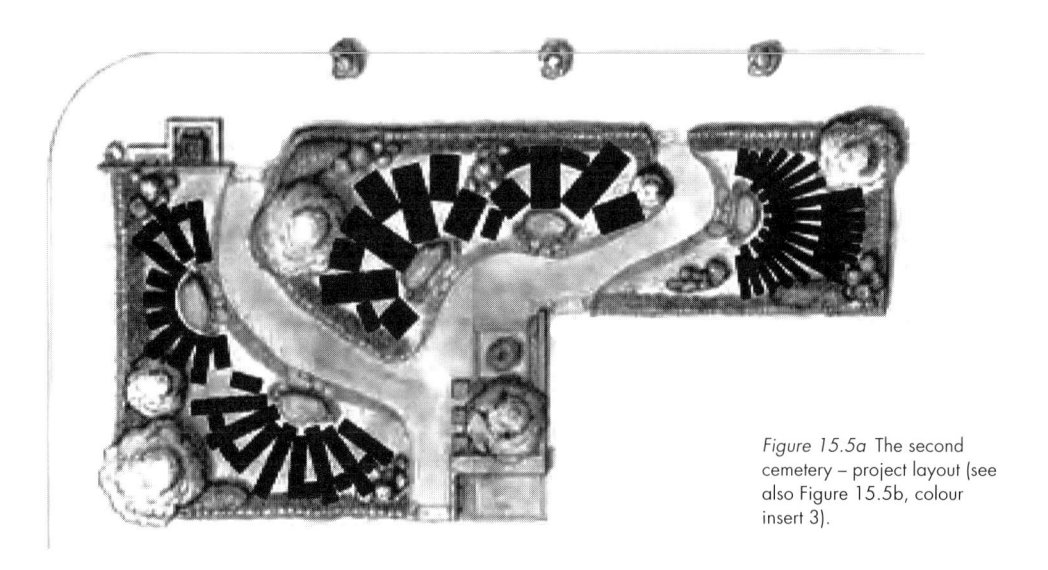

CREATING A CEMETERY

Figure 15.5a The second cemetery – project layout (see also Figure 15.5b, colour insert 3).

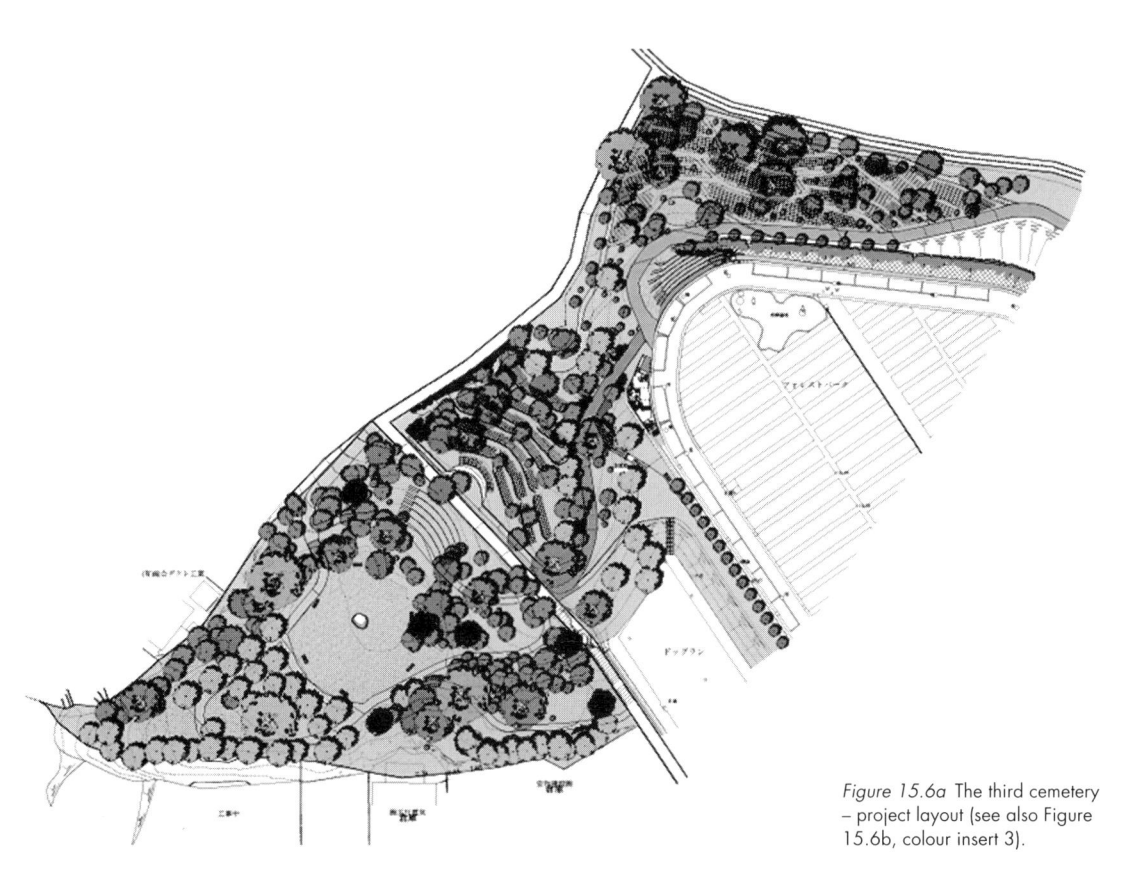

Figure 15.6a The third cemetery – project layout (see also Figure 15.6b, colour insert 3).

The second project is large. It is situated in an area with no existing graves. The concept remains the same, except that if our first solution is a high-density example, the second is an example of a low-density cemetery.

Our third cemetery is surrounded by a small forest. We can say that the cemetery itself is buried in existing nature. We have appropriated part of the forest to make space for 3,000 graves. This project illustrates the point that the Japanese attitude to life and death is complex. Japanese traditional cemeteries tend to reflect part of that attitude in a mixture of Confucian and Buddhist worldviews. A Shinto-based understanding positions life and death within an overall reverence of nature. Our third project closely follows the Shinto sensibility. In terms of design, its extreme simplicity was very demanding. The outcome is radical, but culturally it could fit in anywhere in Japan.

The spatial expression of both life and death looks very similar. From the air our cemeteries resemble Japanese suburbs. By aiming towards new architectural and urban forms we seek a better way of reflecting and fitting into local indigenous culture.

Figure 12.3 Masada (Royal Palace in the Desert) aerial view from the north.

Figure 12.5 Superb example of monumental Nabatean architecture. The monolithic building is dug out of the rock. It served for ceremonial purposes as well as a quarry.

Figure 12.7 Kidron Valley promenade and observation point facing the Temple Mount – Harem E. Sharif.

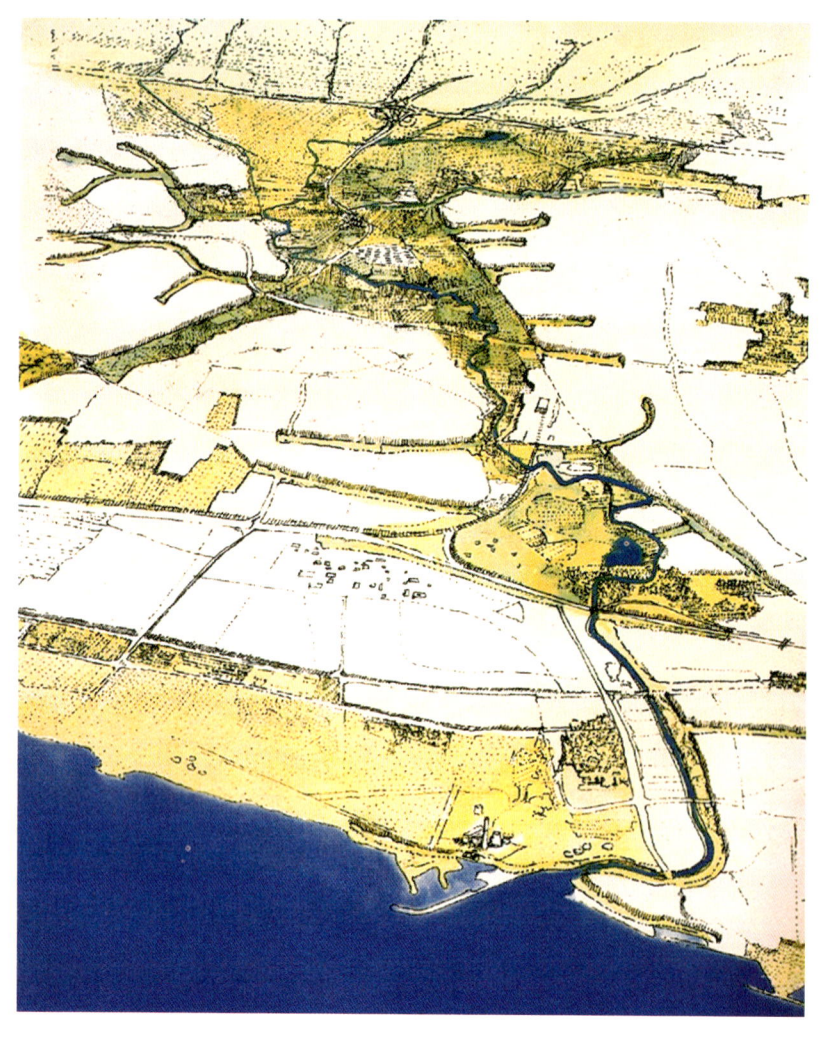

Figure 12.8 The Yarkon River establishes the major green open space corridor with secondary 'green fingers' reaching out to built areas.

Figure 12.10 The Old City of Beersheba – rehabilitation of desert riverbed can catalyse transformation of the whole city and region.

Figure 12.11 In the Old City of Beersheba, the incremental rehabilitation of urban texture – patios, solar envelopes, roof gardens.

Figure 12.12 In the Old City of Beersheba, the large housing neighbourhood and urban block follows democratic distribution of solar rights.

Figure 12.13 Neve Zin – compact housing in desert conditions based on solar envelopes and respect of solar rights of one's neighbour.

Figure 12.14 Neve Zin – social outdoor space also collects run-off water; south-facing windows are critical in cool desert climate.

Figure 12.15 The Albert Katz International School for Desert Studies – main entrance and cooling tower.

Figure 12.16 Centre for Jewish Studies located in this compact building terraced to blend in with Jerusalem topology.

Figure 12.17 The central area of the Centre for Jewish Studies includes an 18 m.-high space for enhancing natural circulation and ventilation, minimizing use of non-renewal resources.

Figure 13.1a One Omotesando Building – façade elements.

Figure 13.1b One Omotesando Building – façade elements.

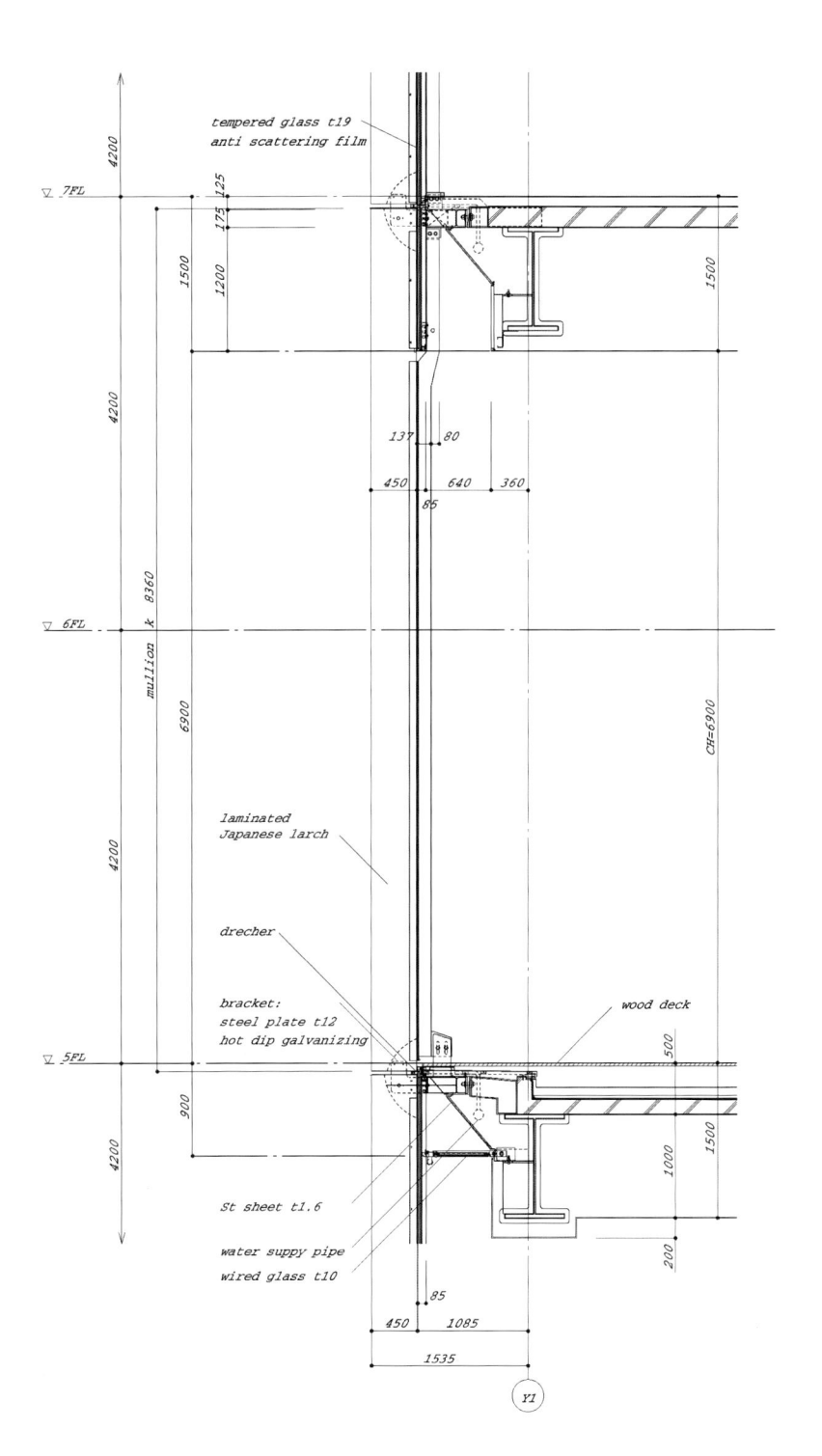

Figure 13.1c One Omotesando Building – façade detail 1.

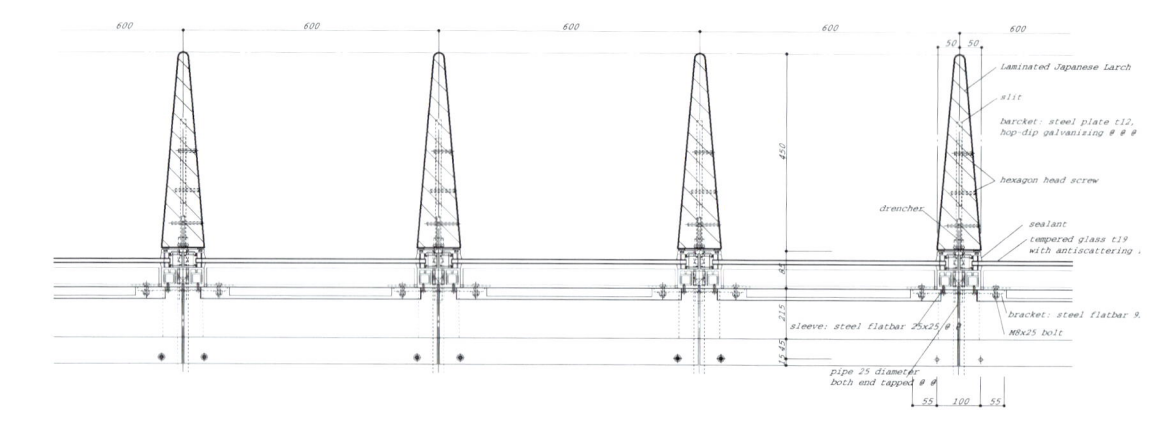

Figure 13.1d One Omotesando Building – façade detail 2.

Figure 13.1e One Omotesando Building – façade.

Figure 13.2a Suntory Museum of Art – façade.

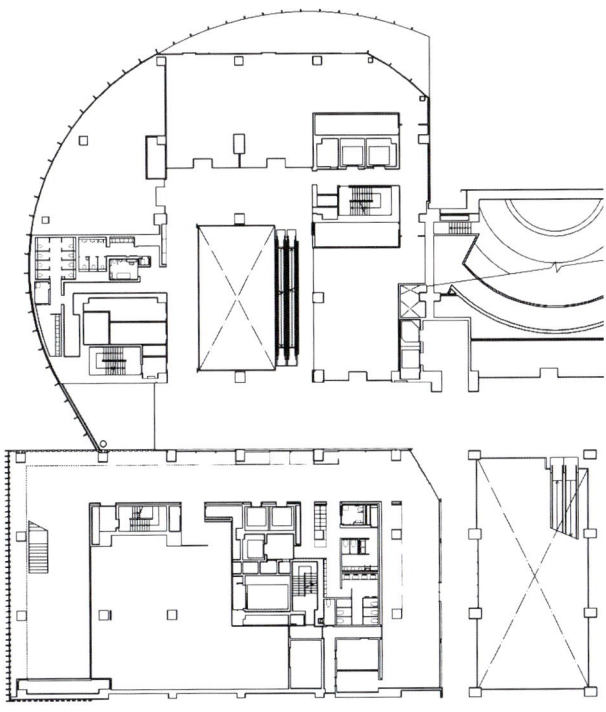

Figure 13.2b Suntory Museum of Art – first-floor plan.

Figure 13.2c Suntory Museum of Art – interior staircase.

Figure 13.2d Suntory Museum of Art – façade sketch.

Figure 13.2e Suntory Museum of Art – façade elements.

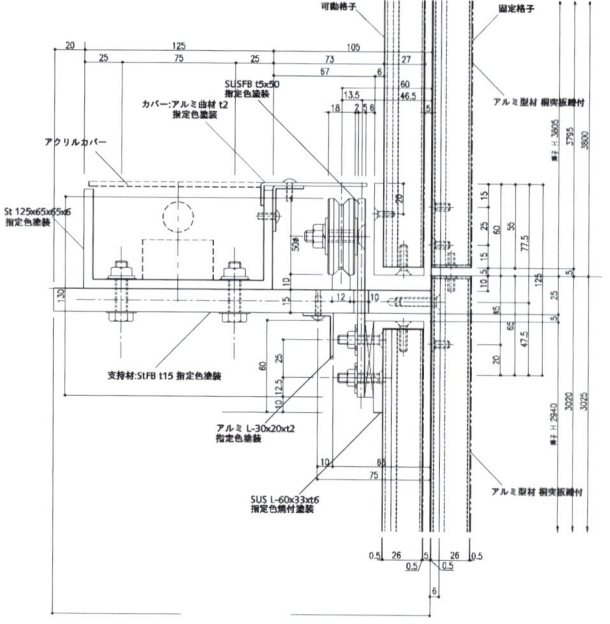

Figure 13.2f Suntory Museum of Art – façade detail 1.

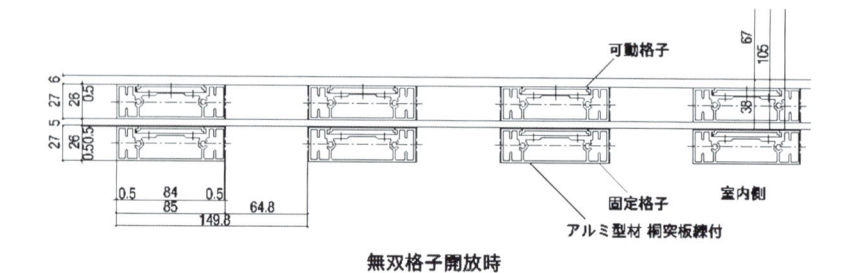

Figure 13.2g Suntory Museum of Art – façade detail 2.

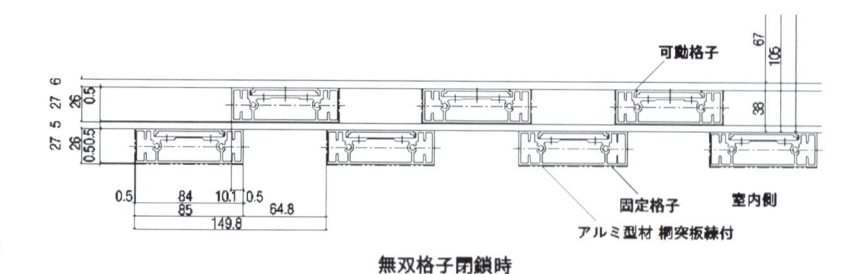

Figure 13.2h Suntory Museum of Art – façade detail 3.

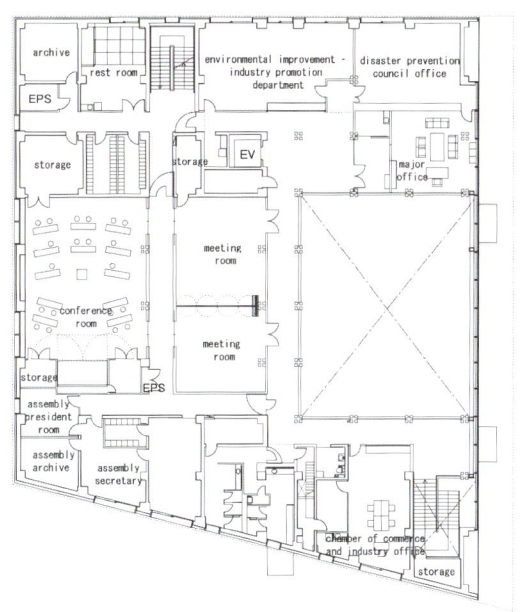

Figure 13.3a Yusuhara Town Hall – ground-floor plan.

Figure 13.3b Yusuhara Town Hall – first-floor plan.

Figure 13.3c Yusuhara Town Hall – façade.

Figure 13.3d Yusuhara Town Hall – façade elements.

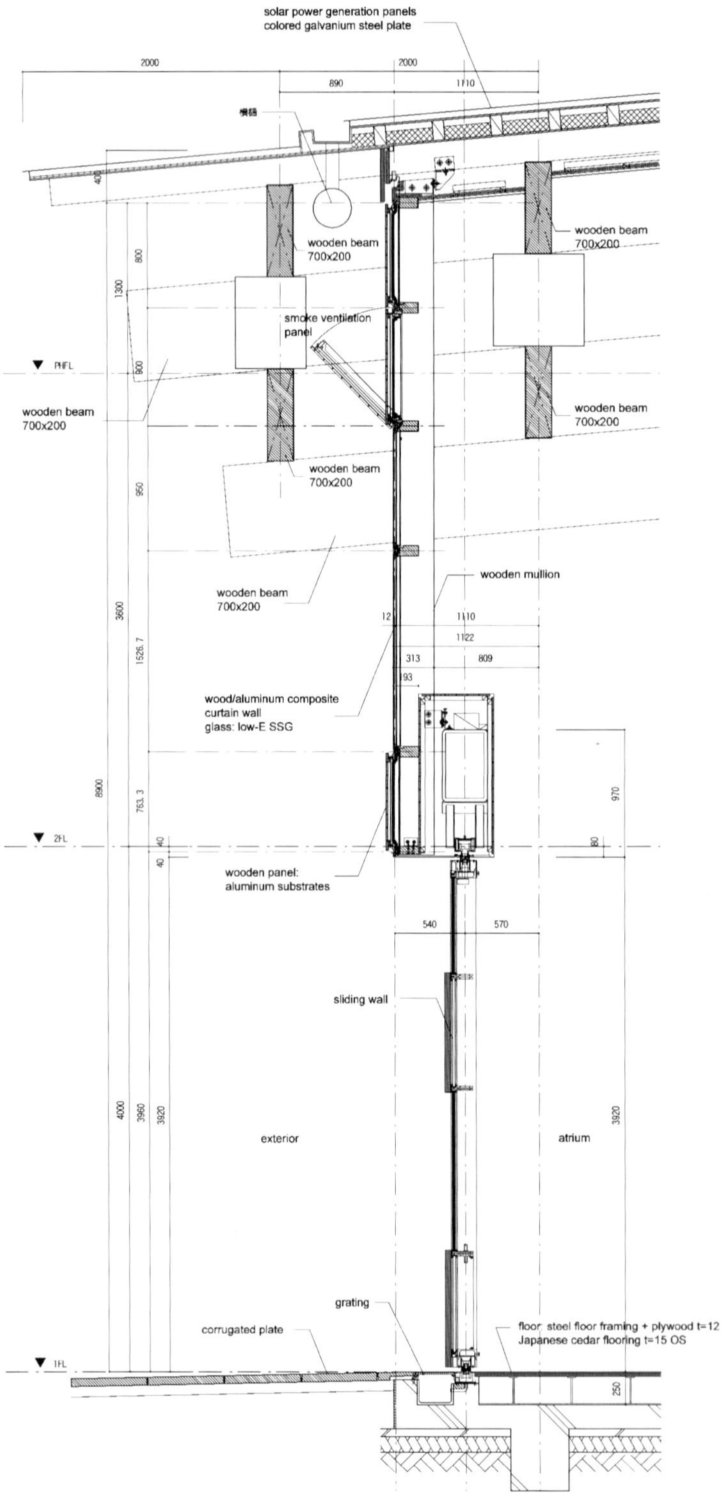

Figure 13.3e Yusuhara Town Hall – façade detail.

Figure 13.3f Yusuhara Town Hall – cross-section.

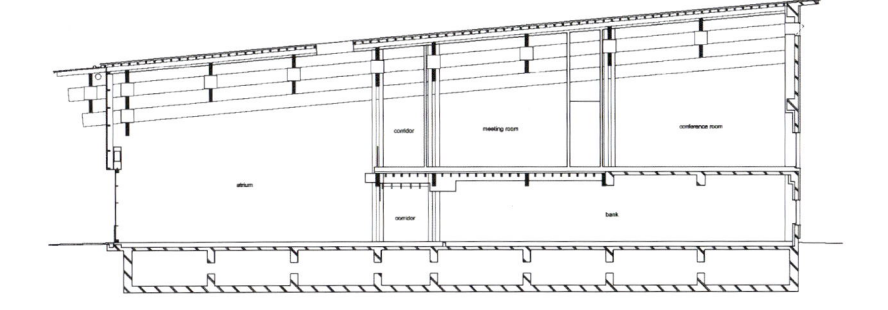

Figure 13.3g Yusuhara Town Hall – semi-outdoor space.

Figure 14.3 Renovation of housing unit for 'Positive Spiral', Osaka.

Figure 14.4a Housing unit – window detail 1.

Figure 14.4b Housing unit – window detail 1.

Figure 15.2a Aerial photograph of a typical residential area in Tokyo.

Figure 15.2b Aerial photograph of a typical cemetery in Tokyo.

Figure 15.2c The typical cemetery in the landscape in Japan.

Figure 15.4b The first cemetery.

Figure 15.5b The second cemetery.

Figure 15.6b The third cemetery.

16 *ECO-URBANITY* HYPOTHESIS
Towards well-mannered built environments

DARKO RADOVIĆ

A PALIMPSEST

During the *eco-urbanity* Symposium, a group of research students from the University of Tokyo undertook to map the key themes, issues and topics that structured the discussion. These terms coalesced around the themes of continuity, scale, shortage, (de)fragmentation, responsibility, the commons, contrasts, design, collaboration and strategies (Figures 16.1, 16.2 and 16.3).

Cultural continuity Continuity
Physical urban continuity
Urban non-urban continuum

Polycentrism + fragmentation
(de)fragmentation
Boundary
Green networks
Scar city
Network

Scale + networks Smallness
Scale Regional, multiscale
Street/urban design
local, urban, regional, national, global

Collective wisdom
Mutual benefit
Ownership
Responsibility
Collectivity Regulation

Shortage
Shortage vs. aspiration
Shortage as catalyst
Smallness

Localizing sustainability

Figure 16.1 eco-urbanity Symposium – a whiteboard summary, session 1.

Community
Public Commons
Common (more than literal meaning)
Empowerment

Coping with urban sprawl
Inevitable expansion
Shrinkage
Contrasts
Life and death Globalization
Closeness/intimacy

Scale
From microscale
Scale/competing scale
Human scale vs. megastructures

Figure 16.2 eco-urbanity Symposium – a whiteboard summary, session 2.

Figure 16.3 eco-urbanity Symposium – a whiteboard summary, session 3.

```
              Materiality                                    Public space
                         Design with nature
         Weak design  Design                                           Collaborations
         Passive design      Sensitivity                     Public advocacy
                  Nature/landscape


                                                          Heritage as resource
         Positive spiral                                     Resources
                       Strategies                    Tradition as critical resource
           Creating collaborations
                              Process of design
         Creating future heritage
                          Responsiveness
         Opportunity from disasters              Scale/small large
                                                    Middle scale    Scale
                                                Walkability    Coexistance
```

At the end of the symposium the participants wrote down the key words that stuck in their minds. Alphabetically presented, their unedited list is shown in Table 16.1.

Overlaid, those two lists form a palimpsest open to diverse readings. If we keep on overlapping terms in this list (constrained only by the limits of a two-

Table 16.1 The resultant tapestry of latent meanings is offered here instead of a conclusion

Coexistence of scales	Spatial editing	Incremental development	Nature	Quality of life, changeable
Compact city	Expanding and shrinking eco-urbanities	Informal, undecided space in (Asian) cities	Participation	Respect for the past
Reconciliation of planned/designed and 'organic'	Formlessness	Landscape	Passive design	Small interventions with big vision
Continuity from the existing circumstances	Green as culture, not as function	Leadership	Poly-centric structures	Soft aspects of the city – collective memory, intangible values, ceremony
Design creativity	Harmony	Learning from the past	Polycentrism	Speed, process of change
Design for shrinking	History	Materiality	Positive (energy) spiral	Super-juxtaposition
Design with nature	Human scale	Multi-scaled, conflicting and complementary eco-urbanities	Public	Sustainable urban form
Eco-urbanity and the 'positive spiral'	Hybrids, coexistence difference	Multiple, heterogeneous eco-urbanities	Public realm	
Tradition as a resource	Transparencies	Urban fragmentation/ de-fragmentation	Urban resistance	Weak design

dimensional page of a book) to these lists generated for, and within eco-urbanity, we will add a list of broad, sustainability-related demands, such as those from the Krockenberger *et al.*'s (2000) definition of sustainability. These are dynamism, people, empowerment, creative potential, quality of life, protecting and enhancing the Earth's life-support systems, variety of life, all people, all generations, fresh air, clean water, healthy soil, protecting nature, fairness. Thus we enrich and deepen possible readings and connotations of the term.

That resultant tapestry of latent meanings is offered here instead of a conclusion.

The reason for this is that I avoid attempts to tame the diversity or to discipline possible readings. Conceptual openness is of special significance to the idea of sustainable development. It refers to a variety of paths towards the ever-changing ideal of the sustainable city discussed in Chapter 1. Rather than presenting fixed definitions and recipes for standardized forms of action, the concept of eco-urbanity asks us to create a field of ideas within which environmentally and culturally responsive and responsible design solutions that represent the best of what is currently available can emerge.

A HYPOTHESIS

At the end of the book, it may be worth going back to the very beginning. The initial invitation to the participants of the *eco-urbanity* Symposium said:

> When we discuss sustainability, we usually speak about ecology. The starting point of our gathering is an understanding that, while rigorous translation of principles of *environmental sustainability* into action is long overdue and remains of critical importance, in times of rampant globalization it is crucial to consider the other side of the dialectical couple which constitutes responsible development – *cultural sustainability*.
>
> Both globalisation and the imperative of sustainable development demand heightened sensibility for the local. It is necessary to develop an ability to judge local social and cultural impacts of imported ideas, practices and technologies. In order to help develop such abilities, the eco-urbanity Symposium will focus on diverse urban environments and local intersections between sustainability and urbanity. An undisputable affinity between those two concepts makes harmony of their spatial expressions a logical aim for the responsible practice of urban planning, urban design, landscape architecture and architecture.
>
> Eco-urbanity will explore cultural belonging and the contextualization of ideas and practices that shape the urban. It will begin with a simple hypothesis:
>
> > Disciplines involved in the production of space need to establish and nurture an awareness that the synergy between ecological and cultural sustainability should never be broken.

POSTSCRIPT

During my investigations of the eco-urbanities of Tokyo in the period from 2006 to 2008, my special focus was on the precincts of Nezu and Yanaka and the rare spaces that survived the infernos of both the 1923 earthquake and the American fire bombing of 1945. Nezu and Yanaka (referred to below as: N.Y., Japan) treasure a precious, vertical connection with Edo, Tokyo of the past. In *Another Tokyo* I recorded how, when

> we enter *roji*, the narrow alleys framed by tiny buildings, we encounter their fine-scale detailing, miniscule urbane gestures that hint, rather than impose, the very specific urbanities of that place. Small pot-plants, for instance, which are present all around Nezu and Yanaka, mark the sensitive, nebulous boundaries between the often overlapping private/not-so-private/not-so-public/and public realms. The sense of intimacy in N.Y., Japan is overwhelming. It sets the tone for experiencing those environments, – whilst functional mix makes sure that, regardless how private some of those spaces might be, they do not forbid interaction. Some of the *roji* are dominated by commercial use. Some are purely residential, but never strictly closed. With their (again small!) gestures, those spaces suggest appropriate modes of behaviour – for the locals and for the visitors alike. The grain and the rhythm of urban fabric in Nezu and Yanaka possess an ability to slow down the movement, to scale the outsider to their own measure, to offer both moments of excitement and flows of everydayness (often as an attraction).
>
> (Radović 2008: 32)

One of the most fascinating findings of that project came up accidentally, from those strong *sensations*. In one of my fieldwork diaries I recorded an encounter with a small *roji* in Yanaka:

> Maybe it is the secrets of my first city,... of Mostar, that I unearth in these nostalgic vibes which I feel in N.Y.... *Avlija* ... my grandparents' frontyard, the main yard, the backyard.... All that was huge (in comparison to the spaces of N.Y.), all that was located at the northernmost reaches of the Mediterranean, where the sounds of Islam got interwoven (knotted together?) with Orthodox and Catholic Christianity and Judaism ... (all so different from N.Y.!) ... but our *avlija* resonated with the same sensibility with which the *roji* abound. Something deep, something quiet, comforting, slow; something human, something gloriously ordinary.
>
> (Radović 2008: 32)

That is where

> the physical and the behavioural ... merge into a distinct atmosphere and create a reality which is very human, tactile, able to stir nostalgia. That is one of those fascinating paradoxes of the urban: by reaching its most

specific, truly local quality, N.Y. Japan provides a deeply universal experience (of the kind which adores any historic urbanity) and thus – by being decidedly local – it transcends the locus and breathes in accord with the complexity of the mondialised world.

(Radović 2008: 32)

Eco-urbanity never separates the measurable and the non-measurable. It aims to grasp the totality of *the urban*, the full complexity of the Lefebvrian *oeuvre*. As the examples of environmentally responsive and responsible practice presented in this book amply show, quantitative and qualitative aspects of spatial quality are densely interwoven, they reinforce or critically question each other.

Eco-urbanity remains an *opera aperta*.

BIBLIOGRAPHY

Krockenberger, K., Kinrade, P. and Thorman, R. (2000) *Natural Advantage: A Blueprint for Sustainable Australia*, Melbourne: Australian Conservation Foundation.

Radović, D. (2008) *Another Tokyo*, Tokyo: The University of Tokyo cSUR and ichii Shobou.

INDEX

accountability 1
Adams, R. 3, 7
ageing population 79, 81, 82, 83
Albert Katz International School for Desert Studies 171, 172
Amsterdam stoop 56
Ando, H. 88

Baker, H. 149
Bangkok *see* urban sprawl (Bangkok and Melbourne comparisons)
Barcelona (reuse, compactness, green): city boundaries, and adjacent municipalities 21; density, La Marina del Prat Vermell plan 26–8; empty space 22, 23, 29; enclaves 23; housing reform 23; infrastructural plans 21; intensity 22; land reuse 19–20, 22–3; mobility, and public transport 22; Olympic Games 1992 20–1; open processes 21–2; public spaces 20, 22, 53; research-based plans 24; strategic green 24, 28–9; strategic planning 19, 21; urban fabric 20–1, 22; urban intensity levels 20; urban projects 20–1; urban structure, persistence of 24–6
Barré, F. 16
Barthes, R. 129–30
Beders, S. 13
Beersheba Old City: master and detailed plans for 170–1; orthogonal grid plan 170; solar envelopes 171; water conservation 171
Beijing (geometries of life and formlessness): built forms, characteristics of 134–5; Chinese revolution 137–8; different scales, co-existence of 132–3; dissolving forms 133–7; ecological ethics 132; empty centre 129, 130; European reflections 125–30; formless geometry of quantity 132–3, 137; modernization 137; natural landscape and built fabric, coexistence of 131; past, separation of present from 137; political institutions, shaping of modern 138; public, rise of 137–8; religious and daily life, expression of 134, 135f, 136f; scale and quantity 131–2; spatial logic 129–30, 132–4; urban design, lack of 125; *see also* space, conceived *versus* lived (Lefebvre)
Berlin 33
Bilbao 33–4, 33f
bio-fuel fiasco, US 10
Bobic, M. 14
Bogotá 34
Bookchin, M. 9
Boulevard St Sulpice, Paris 55
brownfield land, use of 64
Brundtland Commission (1987) 12, 141

cemeteries, Japan: city, as metaphorical human body 182; dead, numbers of in Japan 184f; death and burial, bringing to discussion table 183; death, and cultural belonging 162; first cemetery project layout 184; new trends 183–4; second cemetery project layout 185f, 186; third cemetery project layout 185f, 186; typical 183
Cervero, R. 151
Chandigarh, conventional planned city: circulation, importance of 152; densification, potential for 152–3; land use segregation 152; microclimatic changes 153; sector planning 152

Charoentrakulpeeti, W. 69
climate change, need for consensus 46, 112, 121
climate-responsive design: design process 144, 145; ecological process of 145f; graphical representation of 144f; micro-climates 144–5
Clinton Climate Change initiative 46
Clos, O. 3, 7
CO_2 emissions, reducing 69–70, 79, 83, 87, 178, 179
Commonwealth Games 2006 43
compact city: Beersheba Old City, orthogonal grid plan 171; brownfield land, use of 64; car free housing 64; city-making, lost art of 7; compactness, advantages of 149; concept of 8, 63, 64, 82; as contradiction in terms 65; criteria, as untransferable 63; desert conditions 175; high-density debate 64–6; Jerusalem, compact morphology of 163f, 164–5; liveability 7–8; polycentrism 66–7, 69; sustainability discourse, bias 113; sustainable urban environment policy 63; *see also* Barcelona (reuse, compactness, green)
Copenhagen 34f
creativity 2, 16
cultural sustainability 10, 12, 16

Dale, O.J. 106
de Certeau, M. 14
defragmented city: compact centralities, network of 69, 71f; fragmentation 67–9; public investment, for public good 69–70; sustainable built environment, aspects of 69, 70t
Delhi, eco-city concepts: Gol Market 155f, 156f, 158f; map, site location 155f; natural resources, destruction of 154; Rajendra Nagar 156f, 157f, 158f, 159f; *see also* New Delhi, transit city; Shahjahanabad (Old Delhi), walking indigenous city
density: Delhi 149, 150; environmental aspects, rebalancing 29; fine-grained city 50, 52; high-density debate 64–6; Melbourne 35, 40–2; urban sprawl 118–19; *see also* La Marina del Prat Vermell plan (Barcelona)

desert conditions 167, 175
development controls 39
Dongta 66
Dublin, Temple Bar 33, 35f

eco-settlements 77; conceptual framework 141–2; holistic approach 142; tradition, as repository of knowledge 142–3
eco-urbanity: cultural sustainabliity 10; ecological sustainable development, and urban culture 15–16; environmental responsibility/responsiveness 15–16; ethics, individualization of 9–10; sustainabliity, definitions of 12–14; urbanity, concept of 12–16
ecological footprint 45–6, 143–4
Edo 130, 176
Eilat 172–3
energy: buildings, and energy consumption 179–80; ecological footprint model 143–4; Gurgaon, demand in 151–2; low-energy design principals 180–1; renewable 87–8
extreme conditions, building for: desert conditions 167, 175; Indian context 145–8; Masada 145–8; Nabatean planning 145–8

Fahmi, W. 67
Ferry, L. 16
Fibercity 2050 (shrinkage, designing for): cities, interdependence between 81; compact cities 82–3; disaster prevention strategy 83–4; earthquake disaster prevention measures 83, 85; fire spread prevention measures 83, 84–5; Green Finger 81–3; Green Partition 83–5; Green Web *see* Tokyo Metropolitan Expressway; historical buildings, lack of value placed on 89–90; Ichigaya moat promenade proposal 90; linear interventions 80; *meisho* (public spaces), creating new 88–9; planning, flexible execution of 83–4; production *versus* exchange, focus on 80–1; Shinjuku Imperial Gardens, boundary transformation 90; Urban Wrinkle 88–91; weaving fibres metaphor 80; *see also* population decline
fine-grained city: actual smallness 53–4; Amsterdam stoop 56;

fine-grained city *continued*
  Boulevard St Sulpice, Paris 55;
  cities, attractions of 50, 62;
  density, and sustainability 50, 52;
  dwelling entrances, Spain 59–60;
  fine-grained city 52–4; front
  porches, Tokyo 60; Georgian
  house, forecourt and bridge 57,
  58f; ownership diversity 53; poly-
  functional spaces 53;
  public–private threshold 56; public
  spaces 53; Rue du Faubourg
  Sainte-Honoré, Paris 54; Seine
  (Paris), stone wall along 54, 55f;
  structural diversity 52–3; thick
  thresholds 56–61; verandas,
  Australia and Canada 56–7;
  vestibules (lobbies), Scotland 57,
  58f, 59
fragmentation: rich and poor,
  disparity between 67, 68–9; urban
  regeneration 67–9
Fromm, E. 9

gas emissions, reducing 69–70, 79,
  83, 86, 87, 178, 179
gated communities 67
Gehl, J. 43, 53, 54, 56, 62
Gemzøe 53
geometry, fractal versus Euclidian
  132–3, 137
Georgian house, forecourt and
  bridge 57, 58f
globalization: colonial perspective
  1–2, 11; definitions of 11–12; 'free
  market'-driven 15; homogeneity
  11; ideology of 1–2, 3; and
  mondialization 11–12; progressive
  evolution versus steady decline 11;
  sustainability, environmental and
  cultural 12; 'world-class' global
  cities 114
Goethe, J.W. von 15
Gore, A. 10
Granville Island, Vancouver 52
green-house gas emissions *see* gas
  emissions
Green movements 10
green space *see* Barcelona (reuse,
  compactness, green)
Greenwich Millennium Village 66
Gurgaon, automobile city: energy
  demand 151–2; scattered
  development 151–2; urban growth
  151

Hadid, Z. 117

Harraway, D. 3
height controls 38, 39
Heng, C.K. 4, 76
hierarchical planning 148–9
Highmore, B. 3, 14
Himurja Building, Shimla, India 148
housing: car free housing 64; housing
  conditions 105; housing reform
  23; integration of 25, 26–7; urban
  sprawl 117–18
Huang, P.C.C. 137

Ichigaya moat promenade proposal 90
Inconvenient Truth, An (Gore) 10
Indian context (eco-urbanity):
  Chandigarh, conventional planned
  city 152–3; climate-responsive
  design 144–5; contemporary
  solutions 145–8; Delhi context
  154–9; eco-settlements 141–59;
  energy resource flow, ecological
  footprint model 143–4; extreme
  conditions, building for 145–8;
  future city model 153–4; Gurgaon,
  automobile city 151–2; habitat,
  form of 143, 153; New Delhi,
  transit city 149–51;
  Shahjahanabad (Old Delhi),
  walking indigenous city 148–9;
  sustainable development,
  definition of 141
Ishikawa, M. 4, 75
Israeli perspective (eco-urbanism):
  Albert Katz International School
  for Desert Studies 171, 172;
  Beersheba Old City 170–1; Centre
  for Jewish Studies 172; context
  164–5; cultural principles 164;
  desert conditions 175; eco-urban
  planning principles 169–75; Eilat
  172–3; ethical principles 164;
  historical examples 163, 165–8;
  Jerusalem, compact morphology of
  163f, 164–5; Kidron Valley 169;
  Nazareth quarry transformation
  173–5; Neve Zin 171–2;
  responsive planning 175; river
  planning 169; structural principles
  164; urban planning, and White
  City of Tel Aviv 168–9; water
  scarcity 165; Yarkon River,
  rehabilitation of 169

Jacobs, J. 43, 53, 61
Jenks, M. 3, 8, 63, 66
Jerusalem, compact morphology of
  163f, 164–5

Jianfei, Z. 4, 76
Johnson, C. 148
Jullien, F.l 128–9

Kakamigahara: corridors/cores, designation of 94; diverse commons, creating 99–100; fire damage 94, 96f; Forest Corridor, strategic planning for 94–7; illegal garbage disposal 94, 95f; Kakamigahara City, location of 92–4; Learning Forest, creation of 98, 99f, 100f; lost commons, excavating 76; Meditation Forest, creation of 100, 101f; natural heritage forest 96, 97f, 98f; Old Welfare Centre, conversion of 98, 99f; promenade, creation of 98; River Corridor, and connecting of commons 101–2; sand and gravel excavation 94, 95f; Satoyama woodlands, restoration of 94–7; Urban Corridor, and creation of commons 97–9; urbanization, and destruction of commons 92, 93; vacant land, preservation of 97–8; Water and Green Corridor plan 94, 98; water, pollution of 94; woodlands, disappearance of 93–4
Kaye, B. 105–6
Kidron Valley 169
knowledges, situated 3
Kodama, Y. 4, 162
Krockenberger, K. 13
Krushan, A. 4, 77
Kuma, K. 4, 161–2

La Marina del Prat Vermell plan (Barcelona): 2004–10 Housing Plan, Barcelona City Council 26–7; dense compactness, and recentralization of city 26–8; infrastructural developments, and accessibility 27; land distribution parameters 27–8; urban fabric, and integration of housing 26–7
land reuse 19–20, 22–3
Le Corbusier 152
Lefebvre, H. 1, 16, 76–7
Leh hill council complex, India 145–6
Lewis, M.M.B. 117
liveability 7–8, 33–4, 36
Los Angeles 66
lost commons *see* Kakamigahara, lost commons, excavating

Lovelock, J. 9
Lutyens, E. 149

McGee, T.G. 115
Maclaren, V.W. 117
Malmö 33
Mandelbrot, B.B. 132–3
Mao Zedong 138
Masada 165–6
mass rapid-transit systems: Bangkok 115, 116; New Delhi 151; *see also* transportation systems
materials: local skills/materials 38; natural, restrictions and benefits of 161–2, 176–7
Meadows, D.H. 9
Melbourne (industrial cities to eco-urbanity): Active Frontages policy 41f, 44; arts and cultural programme 42–3; city-improvement agenda 33; conservation controls 39; density 35, 40–2; design philosophy 35, 38; development controls 39; *Grids and Greenery* document 38, 40; height controls 38, 39; heritage assets 38, 39; incremental approach 38; liveability, to sustainability 33–4, 46; local skills/materials 38; local–state government cooperation 33; master plans, area-specific 40; mixed use 42–3; *Places for People* document 38; public realm, high quality 43–5; residential population, restoring to central city 35, 39, 40, 42; retail and events strategies 42; Strategy Plan, City of Melbourne 35, 38; sustainability 42, 45–6; transport infrastructure, and connectivity 43; Turning Basin, restored 37f; urban design strategies 33–4; Yarra River, relationship with 35, 36f; *see also* urban sprawl (Bangkok and Melbourne comparisons)
micro-climates 144–5, 153
mondialization 11–12

Nabatean planning: desert culture 167; extreme conditions, responses to 167; Nabatean farm, reconstruction of 167f; thermal walls 167; town building 167; water collection 167
Naito, H. 4, 162

Nankin Street, Singapore (continuity and departure): China Square area, as transitional zone 109, 110; China Town Conservation Area, proximity to 108–9; Chinese immigration 104–5; Conservation Master Plan 1986 108; cultural development, and Concept Plan (2001) 103–4; floor-to-area ratio (FAR) 104; Fook Hai building 106; history of 104; Hong Lim Complex 106, 111; Housing and Development Board flats 106–7, 108; housing conditions 105; Identity Plan 2002 104; Jackson Plan 1928 104; Land Acquisition Act 1966 106–8; living conditions, overcrowded 105–6; new development/selective conservation approach 109; pedestrianized area 107, 108f; population decentralization policy 106; recession 1980s, effects of 107–8; Rent Control Act 1947 105–6; Suez Canal, opening of 105; sustainable development 103; Urban Renewal Programme 106; urban transformation, and continuity 111; vibrant activity hub 110–11

natural resources, destruction of 79, 154

nature: architecture and nature, uniting in Yusuhara Town Hall 177; bringing back 176; interlocking programmes 177; Midtown Suntory Museum of Art 176–7; natural materials, restrictions and benefits of 161–2, 176–7; One Omotesando Building 176; semi-outdoor spaces 177

Nazareth quarry transformation 173–5

neo-liberal thinking 9

New Delhi, transit city: air pollution 151; congestion 151; hierarchical planning 149–50; mass rapid-transit system, plan for 151; population density 150; radial plan, of city 150; transport modes 150

Newton-Brown, C. 119

Nezu 190

Ohno, H. 4, 75

oil crisis 1970s 9

Old City of Jerusalem 163f, 165

Ortiz, R. 11–12

overcrowding 81–2, 105–6

polycentrism 66–7, 69

population decline: ageing population 79, 81, 82, 83; gas emissions, limiting 79; natural resources, finite 79; overcrowding, resolving 81–2; responding to 81; shrinkage, as unavoidable 79

positive spiral structure: low-energy design principals 180–1; NEST21 project, Osaka 181

Postcode 3000 program, Melbourne 35, 39, 40, 42

production of space 10; creativity and innovation, need for 16; cultural sustainability 16; eco-urbanity value system 16–17; right, to urban sustainability 16; sustainability, and urbanity 16

public spaces: Barcelona 20, 22, 53; fine-grained city 53; Tokyo 88–9

Punjab Energy Development Agency office complex 146–7

Radović, D. 14, 15, 142, 190–1

Raffles, S. 104

Rahamimoff, A. 4, 161

rainfall absorption 173

Rajendra Nagar 156f, 157f, 158f, 159f

Ramage, E.S. 14

Randstad, Netherlands 66–7

Ray, L. 11

Rogers, R. 63, 69

Ross, K. 9

Rue du Faubourg Sainte-Honoré, Paris 54

Sardar Swaran Singh 147

Satoyama woodlands, restoration of 94–7

Saul, J.R. 15

Schumacher, E.F. 9, 15

Seine (Paris), stone wall along 54, 55f

semi-open spaces 177

Shahjahanabad (Old Delhi), walking indigenous city: compactness, advantages of 149; hierarchical planning 148–9; mixed land use 149; population density 149; walking distance, activities at 149; zoned land use 148

Shinjuku Imperial Gardens, boundary transformation 90

Sim, D. 4, 8
Simpson, W.J. 105
Singapore: central precincts 107f; history of 103; street vendors, eradication of 121; *see also* Nankin Street, Singapore
Sintusingha, S. 4, 76, 113, 114, 117
sociocultural diversity, vanishing 10
Soja, E.W. 1
solar energy 39
solar envelopes 171
Sorkin, M. 1
space, conceived versus lived (Lefebvre): binary poles, cultural understandings of 128–9; Chinese urban space, and combining of 127, 128f, 130; city, as epistemological layout 129; Renaissance city 126–7, 128f, 129
strategic green (Barcelona) 24; Ciutadella, historical park 31, 32f; Collserola, part of metropolitan park 30, 30f; density and environmental aspects, rebalancing 29; large open spaces, as part of city 28–9; local characteristics, taking into account 29; Montjuïc, large central park 30–1, 32f; Tres Turons, large empty area 30, 31f
Sun Village, and 'eco lifestyle' 47–8
sustainability, definitions of: complexity of 13–14; cultural sustainability 10, 12, 16, 189; ecological sustainability 9, 12; economic sustainability 12–13; emotional tone of 13; enviromental sustainability 12, 189; ideological struggles in 13; socioeconomic and ecological components 120–1
sustainable design, positive spiral: air-conditioning systems, impact of 178, 180; building evolution, and energy consumption 179–80; gas emissions, lowering of 178; indoor and outdoor environment, relationship between 178, 180; low-energy design, role of 180; positive spiral structure 180–1
Symposium, eco-urbanity, Tokyo: cultural difference 2–3; tapestry of latent meanings 188–9; whiteboard summaries 187–8

Taylor, P.J. 14
Tel Sheba, water collection 168
terminology 4–5
thick thresholds, contributions of: possessions, personalization of space with 60; security, enhanced sense of 60–1; social interaction 60
Thorman, R. 13
Tokyo 65; empty centre 129, 130; map of 88f; natural materials, restrictions and benefits of 176–7; *see also* Fibercity 2050, shrinkage, designing for
Tokyo Metropolitan Expressway: buildings, and energy use 87; earthquake evacuation 85; elevated green corridor, creation of 85–6; new buildings, demand for 88; renewable energy sources 87–8; Tokyo cityscape, transformation of 86, 89, 91
Tokyo Olympics (1964) 86
Torremolinos, concentration of activity in 47f, 48, 49f
transportation systems: accessibility 27; ageing population 83; mass rapid-transit systems 115, 116, 151; poor and well-off, disparities between 68–9; scattered development 151–2

UN Impact Cities for Climate Change 46
urban sprawl (Bangkok and Melbourne comparisons): climate change, need for consensus 112, 121; communications technologies, best practices 113; compact city bias 113; competing interests 112f, 113; cultural foundations 119; eco-urbanity 120–1; economic aspects, bias towards 114; high-density developments, opposition to 118–19; housing 117–18; local/global balance 113, 114; planned expansion 117; ribbon development 115, 116; socioeconomic divisions 115, 116, 118, 119; street-vendor market culture 115, 116, 117, 121; sustanabliity, global challenges of 113, 114, 120–1; transport infrastructure 115–16, 117; urban design 119–20; Western city bias 113
urban structure, persistence of (Barcelona): Cerdá block, as minimum size of intervention 26; existing factories, consideration of

urban structure *continued*
26; existing housing, integration of 25; flexible zoning solutions 26; industrial sector, development of 24; permanent elements, building around 25; public facilities and service spaces 26
urbanity, concept of: globalization process 15; subjectivity 14; urban culture 14–15

verandas, Australia and Canada 56–7
vestibules (lobbies), Scotland 57, 58f, 59
Vollaard, P. 153

Wang, G. 132
Wang, Q. 132
water conservation: Beersheba Old City 171; Jerusalem 165, 163f; Masada 165–6; Nabatean planning 167
Waverly Steps, Edinburgh 51–2
White City of Tel Aviv 168–9
Whitehead, M. 11
Williams, K. 63
Wong, T.C. 105
woodlands *see* Kakamigahara, lost commons, excavating
World Cities 15

Yanaka 190
Yap, A.L.-H. 105
Yarkon River, rehabilitation of 169
Yasmeen, G. 121
Yeoh, B.A. 105

Zijderveld, A. 14
Zimmermann, W. 69
Zizek, S. 1

eBooks – at www.eBookstore.tandf.co.uk

A library at your fingertips!

eBooks are electronic versions of printed books. You can store them on your PC/laptop or browse them online.

They have advantages for anyone needing rapid access to a wide variety of published, copyright information.

eBooks can help your research by enabling you to bookmark chapters, annotate text and use instant searches to find specific words or phrases. Several eBook files would fit on even a small laptop or PDA.

NEW: Save money by eSubscribing: cheap, online access to any eBook for as long as you need it.

Annual subscription packages

We now offer special low-cost bulk subscriptions to packages of eBooks in certain subject areas. These are available to libraries or to individuals.

For more information please contact webmaster.ebooks@tandf.co.uk

We're continually developing the eBook concept, so keep up to date by visiting the website.

www.eBookstore.tandf.co.uk